Two Degrees

The Earth's temperature has been rising. To limit catastrophic outcomes, the international scientific community has set a challenging goal of no more than 2 degrees Celsius (3.6 degrees Fahrenheit) average temperature rise. Economists agree we will save trillions of dollars by acting early. But how do we act successfully? And what's the backup plan if we fall short?

Setting politics aside, *Two Degrees* reviews the current science and explains how we can set practical steps to reduce the extent of warming and to adapt to the inevitable changes, all while improving the bottom line, beautifying our communities, and improving human health. The book is a practical guide intended for a broad audience of those who occupy and shape our built environment. The authors provide a clear framework for communities, policy makers, planners, designers, developers, builders, and operators to help manage the impacts and capture the opportunities of our changing climate.

Two Degrees is divided into three sections—Fundamentals, Mitigation, and Adaptation—covering a diverse array of topics ranging from climate-positive communities and low-carbon buildings to the psychology of choice and the cost of a low-carbon economy. After a foreword by Amory Lovins, more than 10 contributing authors share knowledge based on direct experience in all aspects of built environment practice. This book clarifies the misconceptions, provides new and unique insights, and shows how a better approach to the built environment can increase resilience and positively shape our future.

Alisdair McGregor is a Principal of Arup and Arup Fellow, based in San Francisco. **Cole Roberts** is an Associate Principal of Arup, based in San Francisco. **Fiona Cousins** is a Principal of Arup, based in New York.

Two Degrees

The Built Environment and Our Changing Climate

Alisdair McGregor, Cole Roberts, and Fiona Cousins

Routledge
Taylor & Francis Group

LONDON AND NEW YORK

First published 2013
by Routledge
2 Park Square, Milton Park, Abingdon, Oxon, OX14 4RN

Simultaneously published in the USA and Canada
by Routledge
711 Third Avenue, New York, NY 10017

Routledge is an imprint of the Taylor & Francis Group, an informa business

British Library Cataloguing in Publication Data
A catalogue record for this book is available from the British Library

Library of Congress Cataloging-in-Publication Data
McGregor, Alisdair.
Two degrees : the built environment and our changing climate / Alisdair McGregor,
Cole Roberts, Fiona Cousins.
    p. cm.
Includes bibliographical references.
1. Climate change mitigation. 2. Biodiversity conservation. 3. Ecosystem management.
4. Coastal zone management. I. Roberts, Cole (Stephen Cole) II. Cousins, Fiona.
III. Title.
    QC903.M396 2013
    363.738'74--dc23

ISBN13: 978-0-415-69299-1 (hbk)
ISBN13: 978-0-415-69300-4 (pbk)
ISBN13: 978-0-203-08299-7 (ebk)

Book designed and typeset by Alex Lazarou

Contents

Part 1
Fundamentals   1

Part 2
Mitigation Strategies   55

CONTENTS

Figures and tables

TABLES

Contributors

Alisdair McGregor has over thirty years of experience in the design of buildings for low-energy performance. He has led design teams for a wide variety of Arup projects including hospitals, research labs, corporate offices, museums, and civic facilities, and has a particular interest in the integration of sustainable design principles. He is always looking to find creative solutions that do more with less. As a leader in the field of sustainable design, he is very proactive in searching for environmentally sound solutions to help design intelligent buildings that make as small a demand as possible on the environment and its resources. He lectures frequently at conferences, events, and educational forums on the subject of sustainable design and health care. Alisdair was elected as an Arup Fellow in 2004 in recognition of his contribution to the sustainable design agenda within Arup and the construction industry. He was granted an Honorary Senior Fellowship in 2005 by the Design Futures Council for his noteworthy leadership in the advancement of design solutions. He was named one of Fast Company's Top 100 Creative People in Business in 2011 (at number 17). He leads the Buildings Practice in Arup's San Francisco office.

Cole Roberts leads the energy and resource sustainability business in Arup's San Francisco office. Specializing in design, planning, and consultation in the new and existing built environment, Cole has led dozens of projects to successful LEED certification, including numerous platinum and NZE achievements. Cole has been a keynote speaker at numerous conferences, is a published contributor to peer-reviewed journals, and is a frequent guest lecturer at Stanford University and the University of California at Berkeley.

Fiona Cousins is a principal in the New York office of Arup. She leads the sustainability team, which focuses on translating sustainability aspirations into tangible actions that make a real difference on projects. She also works as a project leader for large, multidisciplinary buildings. Projects include Princeton's Frick Chemistry Laboratory, the High Museum of Art expansion, and the U.S.'s New London Embassy.

Stephanie Glazer advances climate change mitigation measures through planning for sustainable development. An accredited GHG lead verifier in voluntary and compliance markets, she has developed and verified GHG emissions inventories in the telecommunications, power and water/wastewater utilities, waste management, retail, and entertainment industries. Stephanie authored a hybrid production- and consumption-based carbon accounting protocol for development projects and has provided technical advice to the Clinton Climate Initiative's Climate Positive Development Program. She evaluates policy effectiveness and currently participates on ICLEI's Technical Advisory Committee for the Community-Scale GHG Accounting and Reporting Protocol.

Jake Hacker is a building physicist and climate scientist based in the Arup London office. His main area of interest is the mathematical modeling of the performance of low-energy buildings. Jake has a long-standing interest in the science of climate change. He led Arup's Drivers of Change research on climate change, and before joining Arup in 2001 studied environmental fluid dynamics at Cambridge University and worked at the UK government's climate prediction institute, the Hadley Centre for Climate Prediction and Research. Jake is currently a Royal Academy of Engineering Visiting Professor and Senior Teaching Fellow at University College London.

Amy Leitch is a sustainability consultant with Arup, specializing in low-carbon design and climate change resilience building. She has a background in sustainable building design, business strategy, land-use policy, stakeholder engagement, and climate science. Amy earned her master's degree from the Duke University Nicholas School of the Environment and Earth Sciences, with a concentration in environmental law and environmental economics enhanced by program evaluation. Her clients have included the U.S. government, UN-HABITAT, the Bolivian National Park Service, and the C40.

Afaan Naqvi is a mechanical and energy engineer in the Arup San Francisco office, with integrated skills in energy, water, waste, and sustainability. He consults on new and existing building performance and resource efficiency, renewable energy generation, and carbon footprinting. Afaan also has a keen interest in district energy flow and conservation, and has presented on the topic at various conferences.

Simon Roberts is associate director and energy specialist in the Foresight, Innovation and Incubator Group of Arup. Dr. Roberts is a physicist with an industrial background in manufacturing, and with a long-term involvement in sustainability and energy-related matters. His current research focus is on the 4see modeling framework, which combines socioeconomic–energy aspects of a country in order to formulate physically consistent scenarios twenty years or more into the future.

Mark Watts is a director based in London, in Arup's energy consulting team. Focused on cities and sustainability, Mark leads Arup's partnership with the C40 group of the world's largest cities. Prior to joining Arup as a director in November 2008, Mark was a climate change and sustainable transport adviser to the Mayor of London for eight years. Mark led the development of London's groundbreaking Climate Change Action Plan and the associated program of projects to reduce London's carbon emissions by 60 percent by 2025. He was also responsible for leading London's draft climate change adaptation strategy and, on behalf of the Mayor, for establishing the C40 Large Cities Climate Group.

Foreword by Amory B. Lovins

I wrote my first professional papers on climate change in 1968 and on nuclear proliferation in 1973. It was clear then to any thoughtful observer that these were energy's two existential threats to civilization, and that we must take the utmost care to avoid both—and certainly not to trade off one for the other. Remarkably misguided policies have now given us both. Yet now as in the 1960s we can still greatly mitigate both threats by choosing the best buys first, rather than the worst. To be sure, many designers of efficient buildings and communities now seek climate-adaptive designs too, either as a hedge against slow adoption of known solutions for mitigating climate change or because some climate change is now unavoidable. But often similar design approaches, as this book describes, can achieve both mitigation and adaptation goals together, at or below normal construction cost.

The best of the best buys is radical energy efficiency, most of all in buildings and communities. Globally, 40 percent of fossil carbon emissions are due to buildings and 24 percent due to transport between buildings; the rest are from industry. Most of this energy is wasted, and technologies for using it more productively keep improving faster than they're installed, so the unbought "efficiency resource" keeps getting bigger and cheaper. U.S. buildings, for example, could save $1.4 trillion in net present value by wringing three to four times more work from their energy. The savings would be worth four times their cost, earning an average 33 percent internal rate of return (not counting nonenergy benefits that are often much larger). Industry, too, could double its energy productivity, with a 21 percent internal rate of return.

These empirically grounded, peer-reviewed findings rely most of all—especially in our built environment—on a quiet revolution called "integrated design" that this valuable book helps to explicate (in Chapter 10 and throughout). Optimizing whole systems for multiple benefits, rather than isolated components for single benefits, can often make very large energy savings cost less than small or no savings, turning diminishing returns into expanding returns. This is true not only of energy but of all resources, and not just in buildings but across whole urban systems and their resource sheds—transportation, water, waste, materials, and carbon.

For example, consider water, which is intimately linked to energy in many ways and in both directions. Stanford University's Y2E2 building saves six times as much water indirectly through its energy efficiency as it saves directly through fixtures: synergy beats efficiency. (The building's direct energy savings, increased by higher occupancy and longer hours because everyone wanted to be there, paid back in just two years, a 35 percent annual return on investment.) Similarly, in Pittsburgh's Nine Mile Run, a costly combined sewer–overflow problem (one of more than a trillion dollars' worth across the United States) could be resolved at roughly zero or negative cost through smart landscape architecture and real estate models.

In South Central Los Angeles in 1998, TreePeople experimentally retrofitted a 1920s bungalow so effectively that a thousand-year cloudburst—71 centimeters (28 inches) of rain in 20 minutes, all from fire hoses—stayed entirely on-site. Replicated citywide, this approach could cut water imports 50 to 60 percent, help control flooding, reduce toxic runoff into the sea, improve air and water quality, save energy, cut by 30 percent the flow of yard wastes to landfills (creating instead water-retaining, soil-building mulches and composts), beautify neighborhoods, and create perhaps 50,000 jobs for urban watershed managers. The city's two biggest water agencies, one annually spending a billion dollars bringing water in and the other a half billion dollars to take it away, hadn't previously talked to each other.

The water-and-landscape story is even bigger. Professor Malin Falkenmark in Stockholm notes that the unused "green water" falling on the landscape is so much larger than the "blue water" captured in pipes that retaining and more productively using even a small fraction of the former can dwarf the major efficiency gains available from the latter. Better management of green water could lift billions out of water poverty and make food production far more resilient to climate change.

Or consider the ubiquitous asphalt car parks outside big-box stores and shopping malls. Their dark color absorbs solar heat, baking the store, customers, and their cars in hot air and radiant heat, and reducing paving life. It also absorbs light at night, boosting security-driven night-lighting standards until customers driving off the lot are so dazzled they'll have accidents anyway. In contrast, light-colored paving keeps the store, customers, and cars comfortably cool; makes the paving last almost indefinitely; and by increasing optical reflectance, can cut lighting energy to just 1 percent of official norms with superior visibility and aesthetics. And while repaving the parking area one might as well consider photovoltaic shades and, for that matter, pervious surfaces—which in turn can shrink or eliminate costly stormwater-management infrastructure.

However many resources it encompasses, integrated design of whole systems often starts with energy. That's not the only good handle to grasp: as Chapter 8 of this book describes (and as Chapter 14 of *Natural Capitalism* described in 1999), Curitiba, Brazil, achieved astonishing results through integrative design that was first established with water, food, wastes, and other factors as priorities, then expanded, optimized, and maximized over forty years.

The needed questions are often arrestingly simple. For example, how much thermal insulation should you install in your house in a cold climate? The textbooks say to add only as much as will repay its extra cost from saved fuel over the years. This is methodologically incorrect, because it omits the avoidable capital cost of the heating equipment. My own house—at an elevation of 2,200 meters (7,100 feet) in the Rocky Mountains in Western Colorado, where temperatures used to go down to –44°C (–47°F)—was optimized for both operating and capital costs together (doubling its insulation) and is now simultaneously ripening its thirty-seventh through forty-first passive-solar banana crops with no furnace. When completed in 1983–4, this building used 1 percent the normal space- and water-heating energy, about 10 percent the normal household electricity, and half the water, yet repaid its 1 percent extra construction cost in 10 months. Eliminating the heating equipment saved about $1,100 more in construction cost than it added for superinsulation, superwindows, airtightness, and ventilation heat recovery.

That building helped inspire 32,000 European "passive buildings" that have and need no heating equipment, yet provide superior comfort with comparable or sometimes slightly lower construction cost. In the 1980s and 1990s, the same approach eliminated space-cooling equipment up to 46°C (115°F) in dry climates—not an upper limit—and cut a Bangkok home's air-conditioning energy by 90 percent, in both cases with better comfort and normal or lower construction cost. In all these cases, today's technologies are even better and cheaper. My house has retrofitted them and is measuring their performance—complicated by the annoying tendency of the monitoring equipment to use more energy than the lights and appliances it's measuring.

Similar techniques apply to big buildings too. A few years ago, RMI co-led with Johnson Controls the conceptual and schematic design of an integrated advanced energy retrofit as part of the $0.5 billion renovation of the Empire State Building. Normal checklists of incremental measures were initially proposed to save less than 10 percent of the energy, yet integrated savings achieved over 40 percent. Remanufacturing the 6,514 clear double-glazed windows on-site into superwindows that passed light but blocked heat—cutting winter heat losses by at least two-thirds and summer heat gains by half—combined with lighting and plug-load improvements to cut the peak cooling load by one-third. This allowed the retrofit team to renovate and shrink the old chilled-water equipment rather than replacing and expanding it, saving $17 million in capital cost. This cut the total retrofit cost to $13 million and the payback to three years, with stunning economic advantage to the owner, Tony Malkin, who is spreading this approach to the whole industry.

The late Greg Franta, FAIA and I designed a similarly surprising retrofit for a curtain wall office tower near Chicago in 1994, saving three-fourths of its energy with slightly less investment than its scheduled renovation, which saved nothing. Instructively, the design, though approved by the owner, was not executed. A leasing broker, short of cash and incentivized on deal flow, scuttled the retrofit to avoid delaying commissions. The property then couldn't be re-leased because of poor comfort and high gross occupancy cost, so it was sold off at a distressed price. This illustrates why successful integrative retrofit requires meticulous attention to detail—each of the two dozen parties in the commercial real estate value chain can be a showstopper, though each is a business opportunity—and mindfulness of each party's remarkably perverse incentives. But with trillions of dollars of net present value on the table, there's plenty of reason to pay attention. And today's techniques are even better—permitting, for example, an expected energy saving around 70 percent in the General Services Administration's flagship retrofit, with RMI, of the Byron Rogers Federal Building in Denver, all within federal investment guidelines. That could make it the most efficient office building in the United States despite its poor orientation, 1964 vintage, and requirements for full asbestos abatement, federal blast-resistance retrofit, and historic preservation.

Everywhere, innovation is oozing if not gushing up through the cracks. Walmart's purchasing power drove innovation that cut the cost of radiant floor-slab cooling by 69 percent (see Chapter 14). Nearly 5,000 actions by the mayors of the world's 40 largest cities are uplifting citizens, saving money, and cutting emissions (see Chapter 8). Stanford University expects to save over a fourth of its climate-harming emissions and 18 percent of its water just by integrating its buildings' heating and cooling needs in a campus-wide system (see Chapter 8). Consistently, integrative thinking trumps reductionism.

In short, this book reviews the fundamentals and opens a cornucopia of creative ideas for doing more and better with far less for longer, with lower cost and risk. To be sure, the cornucopia is the manual model—you must actually go turn the crank—but these gifted practitioners point the way to astonishing opportunities.

A concluding word about the organization of this book: Chapter 1 is an enjoyable review of climate science. It is also an update on what has happened since the last IPCC report in 2007. It brings bad news. It may depress you, and it's hard to depress people into action. So if you want to review climate science fundamentals and understand how and why our species has a serious problem caused by experimenting with the planet's climate, read it. But if you already know there's a climate problem and just want to get on with solving it—or if you want to do the same things anyway (whether you believe the climate science or not) for other reasons, such as making money or improving national security—then you can skip straight to Chapter 2. In that and the rest of this excellent book, you can learn to create abundance by design, through practical transformation, in a spirit of applied hope.

Amory B. Lovins
Chairman and Chief Scientist
Rocky Mountain Institute
Old Snowmass, Colorado
March 18, 2012

AUTHORS' NOTE: *We invited Amory to author the foreword due to our deep respect for his and Rocky Mountain Institute's contribution to a sustainable built environment. Founded thirty years ago, RMI continues to be an independent, entrepreneurial, nonprofit think-and-do tank that drives the efficient and restorative use of resources. Its practices in the built environment, transport, industry, and electricity have led to such game-changing publications as* Natural Capitalism *(www.natcap.org),* Reinventing Fire *(www.reinventingfire.com), www.retrofitdepot.org, and hundreds of papers free at www.rmi.org, all complemented by the emerging initiative "10×E: Factor Ten Engineering."*

For those keen on further reading, we recommend RMI's Reinventing Fire *(2011), which shows how a 2050 U.S. economy 2.6 times today's could need no oil, no coal, no nuclear energy, and one-third less natural gas than now; emit 82 to 86 percent less fossil carbon; cost $5 trillion less (in net present value, counting no externalities); need no new inventions or acts of congress; and be led by business for profit.*

Preface

I decided not to tell lies in verse. Not to feign any emotion that I did not feel; not to pretend to believe in optimism or pessimism, or unreversible progress; not to say anything because it was popular, or generally accepted, or fashionable in intellectual circles, unless I myself believed it; and not to believe easily.

Robinson Jeffers (1887–1962), poet

There are many books on theory. This is not one. This book is about getting to a better place, stepping aside from the politics of climate change, and stepping into the practice of creating buildings, infrastructure, and communities that will last us into a human-stewarded future. All of the authors are practicing professionals from diverse backgrounds who have worked on some of the most well-known (and some not so well-known) high-performance projects in the world. They've seen some success and some more valuable failure. And their experience is blessed by the critical comments and prior work of their peer community of "doers"—designers, planners, policy makers, business professionals, appointed and elected officials, builders, farmers, and everyday citizens.

If I were a reader, what would capture my attention, impress and intrigue me? What are my expectations, and what would I like to discover in this book? I will not say opportunity or optimism. These words are too easy to throw around and often difficult to back up. Oversold and under-realized, one might say. A critique of either likely labels the critic a recalcitrant, or worse, a pessimist. I prefer to start at the origin of opportunity—a "coming toward a port"—and the suggestion of safe harbor. This image evokes both where we are and where we need to be. That is, dark seas with a storm brewing—and calm waters ahead, if we can reach them. However, this book is less about the storm and the safe harbor than the sails, oars, and spirit that separate the two. To better set the scene, I pull two words from the following pages, the word *synergy* and the word *compete*. The two define this book, because they tell us how to reach our goal, not just how nice it will be when we arrive. Their meaning is closer than one would expect: *synergy*, "to work together," and *compete*, "to strive together."

Our companies, countries, and people certainly do strive, but do they realize they strive together? That they push each other toward a higher *shared* goal?

Historically, the solution to shortage and growth has been efficiency. Less use per unit output. More comfort per unit input. But what if the whole idea of efficiency is incomplete? Like a fundamental law of physics, it makes sense, but only until you look outside the car and realize that it is driving toward a cliff, very efficiently.

Among the authors there is over 110 years of experience. The reviewers triple that. But more important is the range in years of experience, gender, origin, education, and age. A single voice often focuses on efficiency, but diversity seeks synergy and fosters competition. Although any of us could have written a book with a similar title (and others have), I don't believe any of us could have written this book alone.

Within its chapters, the book reviews many topics in the built environment. It lays out a methodical path for lowering carbon emissions toward zero and moving toward climate-positive communities, while simultaneously making money from the investments and creating healthier places for our families. It describes a proven process for building resilient communities that are not hobbled by extreme weather and the domino effect of accumulated risks. It showcases examples of success and failure among cities, home builders, major institutions, and large corporations.

The questions this book answers are the questions many of us ask. Can we afford a low-carbon economy? How do the trillions of square feet of buildings and billions of people *really* change—not just how *should* they change? How can carbon neutrality be both possible *and* profitable? Why do we make the decisions we do, and can we work *with* our irrational tendencies and default decisions? If we do need to adapt, what will adaptation really feel like?

In writing this book, we aimed to answer questions that each of us had in our heart. We also wanted to share our learned insights so that the ideas can go farther and wider than each of us can physically. That is, we wanted to both gain knowledge and share it.

Alisdair's windsurfing brings him often to the beach. There on the doorstep of the Pacific, he tells me he can see the ocean of his children's old age: nearly 2 meters (6.5 feet) higher and over his head by recently revised estimates, the sand beneath his toes long since washed away and the nearby buildings either flooded or barricaded behind seawalls. Fiona's hometowns of New York City and London are just as threatened by rising seas. When I visit my family's farms and ranches in North Dakota, Minnesota, and Wyoming, I wonder how such distant effects will reach them. Will farms fail when reefs do? It's hard not to worry in general, and not to be confused by debate. The impacts are also easy to ignore on a sunny day among friends and family. Would it be as easy to ignore an impending asteroid strike, an "impact event" that squeezes all the change of a century into just a few years?

When Alisdair first outlined this book nearly four years ago, the content was much as you'll read it today. It is a play in three parts—Fundamentals, Mitigation, and Adaptation—with an emergent voice from its many authors. Chapters are (1) reviews of fundamental science, (2) position chapters that carry the weight of their authors' opinions, or (3) guideline chapters that are intended to convert easily to the professional and personal practice of the reader. References and a glossary are provided for further reading and corroboration.

The bulk of the insight and inspiration—arguably the credit for this book as a whole—stems from the projects, people, collaborators, and competitors of Arup. Although the authors work or have worked for Arup, this is not a book by a firm. The time has been largely our own, as will be the responsibility for error and omission.

We hope you will open the pages of this book and find it valuable now and in the years ahead—in terms of creativity, methodology, and how easy and how hard practice is compared to theory. We also hope that you will be inspired to continue the conversation among friends, family, and colleagues with passion and shared experience. Only through synergy and competition, working together and striving together, will we arrive safely in our harbor.

Cole Roberts, PE, coauthor
Twenty-two years after 1990 baseline emissions (2012)

Acknowledgments

This book is formed from the hard work of tens of thousands of individuals around the world, each striving to improve understanding of climate change, form effective policy, and shape better buildings, cities, and infrastructure. In addition to the emergent voice from all these people, the authors would like to acknowledge the insights and inspiration provided by the following people and organizations:

From the Arup community Engin Ayaz, Stephen Belcher, David Brown, Stephen Cook, Adam Courtney, Jo da Silva, Raj Daswani, Steve Done, Martín Fernández de Córdova, Laura Frost, Karin Giefer, Stephanie Glazer, Jake Hacker, Chris Jofeh, Sam Kernaghan, Amit Khanna, Amy Leitch, Chris Luebkeman, Afaan Naqvi, Jordan O'Brien, Tim Pattinson, John Roberts, Simon Roberts, Davina Rooney, Christopher Rush, Jeffrey Schwane, Pauline Shirley, Robert Stava, Mike Sweeney, Cameron Talbot-Stern, Vinh Tran, Polly Turton, John Turzynski, Chris Twinn, Mark Watts, and Frances Yang.

And most of all, Jesse Vernon, for her editing skill and commitment to quality.

From the design community Clark Brockman (Sera Architects), Jeb Brugmann (ICLEI), Margaret Castillo (Helpern Architects), Judy Corbett (LGC), Kaitlyn DiGangi (consultant), Rosamond Fletcher (AIA NY), Robert Goodwin (Perkins & Will), Dave Johnson and William McDonough (WMP), Amory Lovins (RMI), Huck Rorick (Groundwork Institute), Bry Sarte (Sherwood), Severn Suzuki (ECO), Sim Van der Ryn (former CA State Architect), and Donna Zimmerman.

From the policy community Sam Adams (former mayor of Portland); Keith Bergthold (City of Fresno); Cal Broomhead, Gail Brownell, and Rueben Schwartz (SF Environment); Center for Climate and Energy Solutions; the Institute for Social and Environmental Transition; Ken Livingston (former Mayor of London); Heidi Nutters (BCDC); the Rockefeller Foundation; and the team at the World Bank Climate Change program.

From the research and climate science community Nancy Carlisle, Shanti Pless, and Otto Van Geet (NREL); Jared Diamond (UCLA); B.J. Fogg (Stanford University); the Intergovernmental Panel on Climate Change; Geoff Jenkins (UK Hadley Centre); Ron Prinn (MIT); Chris Sabine (NOAA); and Professor Keith Shine (Reading University).

Our friends and families We are grateful for the insights, ad hoc reviews, and encouragement of Mackenzie Bergstrom, Olly Gotel, Kash Heitkamp, Jenny McGregor, Raymond Quinn, Hayes Slade, and Serena Unger.

PART 1

Fundamentals

A view of our atmosphere at high altitude.
From up here, it looks fine. Can we keep it that way?
Image: MarcelClemens/Shutterstock.com

1

The Science of Climate Change

Jake Hacker and Cole Roberts

> There are in fact two things, science and opinion; the former
> begets knowledge, the latter ignorance.
>
> Hippocrates (460–377 BCE)

Since the mid-nineteenth century, the Earth's average surface temperature[1] has increased by around 0.8°C (1.4°F) (Figure 1.1).[2] Most of this warming has occurred over the last three decades, with the last decade being the warmest decade and 2010 the warmest year since 1850, the start of the global instrumental temperature record.[3] The heat content of the oceans has also increased, and there have been other changes directly associated with a warmer world, notably a retreat of land glaciers and a reduction in the extent of sea ice.[4] These trends are now viewed to be irrefutable, and although there are some regional variations, the whole globe is getting, on average, steadily warmer. This is the phenomenon of *global warming*. There have been other changes that can also be attributed to a warming world, such as changes in rainfall patterns, sea-level rise, and an increase in the frequency of extreme climate events. These changes fall under the broader heading of *climate*

change. The most likely explanation is the impact of human activities on the natural environment, principally a change in the chemical composition of the atmosphere.

The issue of human-made (*anthropogenic*) climate change can be succinctly described by the following two statements:

1. The Earth's average temperature is rising due to an increase in the concentration of greenhouse gases (GHGs) in the atmosphere resulting from human activity, principally the burning of fossil fuels and the clearing of forests.
2. If humans continue development based on fossil fuel energy, global warming and associated climate changes will accelerate, with dangerous consequences for both the human and natural world.

The two parts of the hypothesis are both recognized by the national science academies of all the major industrialized nations to be *true beyond reasonable doubt*.[5] Despite this, there is still a widespread lack of acceptance of their validity, and a lack of real action to reduce

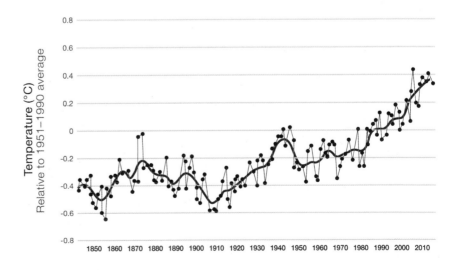

1.1 Global average land and sea surface temperature from direct thermometer measurements since 1850
Graphic: Arup, Data: Met Office (see note 3)

Legend
● Annual data values
━━━ Trend line "decadally smoothed"

emissions of GHGs or to manage the consequences of climate change. In this chapter, then, we will treat these statements as only a hypothesis, which we will call the Global Warming Hypothesis (GWH).[6] We will present the supporting evidence for the hypothesis and go on to discuss some of the implications.

DEVELOPMENT OF THE GLOBAL WARMING HYPOTHESIS

The GWH is not a new concept. It was first proposed by Nobel laureate Svante Arrhenius at the turn of the twentieth century. Arrhenius was fascinated by the problem of explaining the ice ages. In 1896, he published a paper proposing that past climate changes may have been produced by alterations in atmospheric carbon dioxide (CO_2) levels. Through hand calculations and building on earlier work by Fourier, Tyndall, and other nineteenth-century scientific pioneers, he showed that a doubling of atmospheric CO_2 could have produced an increase in global average temperature of around 4°C (7°F)—enough to account for an ice age.[7]

However, Arrhenius's most prescient step came a few years later. He argued that the use of fossil fuels could lead to an increase in atmospheric CO_2 and that doubling the concentration would produce global warming of around 5 to 6°C (9 to 11°F). He was not overly concerned by this prospect.[8] Indeed, as a resident of Sweden, he offered that the warming would make the world a more amenable place for humans to live.

The GWH was recast in a more ominous form in the late 1950s, by scientists concerned about the accelerating use of fossil fuels. Even then, most scientists thought the CO_2 released into the atmosphere would be absorbed into the Earth's natural stores of carbon—the world's forests and oceans—but in 1957 two prominent U.S. climate scientists, Roger Revelle and Hans Suess, published a paper showing that the CO_2 produced from burning fossil fuels was building up in the atmosphere and that the oceans had much less capacity to absorb CO_2 than previously thought.[9] These findings started the modern investigation of the GWH, involving a combination of measurement and theoretical investigations. The emerging findings caused the Geneva World Climate Conference in 1979 to conclude: "It is highly credible that a doubling of CO_2 in the atmosphere will bring 1.5–4.5°C [2.7–8.1°F] global warming."[10]

After thirty years of further investigation, this range of possible temperature change for a doubling of atmospheric CO_2 has remained largely unchanged. However, there has been a significant amount of effort to refine the calculations and to reduce uncertainty. In 1979, two major areas of uncertainty remained: first, there was not yet a firm observational confirmation of a global warming trend, and second, while the physical effects of changes in atmospheric CO_2 on climate were well understood, other factors that can also influence climate were more uncertain.

FACTORS THAT DETERMINE CLIMATE

Climate is time-averaged weather. The factors that determine day-to-day weather are complex and, as you know if you've ever been frustrated by the weather forecast, evidently difficult to predict. However, the factors affecting climate are stable and easy to understand.

The temperature of a planet's surface is determined primarily by two factors: the amount of sunlight that reaches the surface and the rate at which heat is radiated away from the surface into space. A crucial aspect of the physics of climate change is that the thermal radiation carrying these two heat flows occurs at different wavelengths in the

TABLE 1.1

Analogy between the Earth's climate and a room heated only by the Sun
Data: J. N. Hacker, "Climate as a Driver of Change: Part 1: The Evidence and the Causes," *Arup Journal*, 2007, 1.

Earth climate	Room climate
Sun	Sun
temperature of lower atmosphere/ Earth's surface	room air temperature
greenhouse effect	insulation
sea and land ice	reflective blind
clouds	curtains
ocean	thermal mass of building fabric

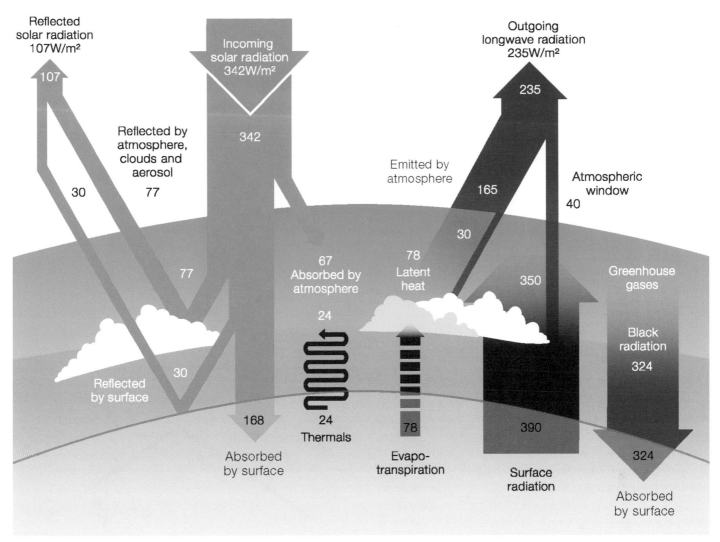

1.2 The Earth's radiation balance

Graphic: Arup, adapted from Kiehl and Trenberth, 1997

electromagnetic spectrum. This is because the surfaces of the Sun and the Earth are at very different temperatures. The Sun produces shortwave radiation in the visible and near-visible parts of the spectrum, whereas the Earth produces longwave infrared radiation. In thermal equilibrium, the two heat flows are equal. The warmer the Earth is, the more infrared radiation it produces, so equilibrium is reached when the Earth's surface is at just the right temperature to balance the shortwave radiation received from the Sun. Another way to think about this is that the Earth's atmosphere is like the inside of a building built of glass (e.g., a greenhouse). The atmosphere is the glass; the land and oceans are the floor; the near-surface air temperature is the air temperature in the building (Table 1.1).

The amount of sunlight reaching the surface is determined by the Earth's "shading": the reflection of sunlight from the tops of clouds

and from snow and ice. About 30 percent of the Sun's radiation is reflected directly back into space, 25 percent from clouds and 5 percent from ice (Figure 1.2). Shading is also provided by atmospheric aerosols—fine particles such as smoke, fog, and mist. Aerosols have natural sources, such as volcanoes, but there are also anthropogenic sources, principally from burning biomass and fossil fuels. Light-colored aerosols reflect sunlight, so they have a cooling effect, but dark-colored aerosols, such as soot, absorb sunlight, so they have a warming effect. Aerosols also promote cloud formation, which increases shading, so they have an indirect cooling effect.

The Earth's "insulation" is the atmosphere. The atmosphere is largely transparent to shortwave radiation but mostly opaque to infrared radiation because of the GHGs it contains, which effectively reflect some of the longwave radiation back toward the Earth's surface (Figure

TABLE 1.2
The main GHGs in the atmosphere today, with approximate contributions to the elevation of the Earth's surface temperature.
Data: R. G. Barry and R. J. Chorley, *Atmosphere, Weather and Climate*, 7th ed. (London: Routledge, 1998)

Greenhouse gas	Contribution to temperature changes
water	+21°C
CO_2	+7°C
O_3	+2°C
others (CH_4, N_2O, CFCs)	+3°C
Total	+33°C

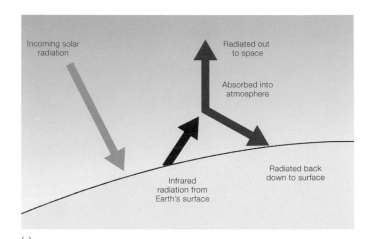

(a)

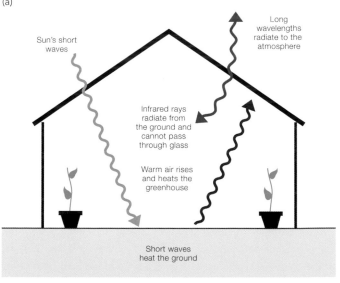

(b)

1.3 The greenhouse effect: (a) Earth, (b) greenhouse
Graphic: Arup

1.2). This means that the Earth must produce more radiation—i.e., get warmer—to transmit the same net radiation through the atmosphere out into space. Without the GHGs, the Earth's surface would be around 33°C (59°F) cooler (Table 1.2). This effect is called the *greenhouse effect* because glass is also largely opaque to longwave radiation (Figure 1.3). In window systems, the effect can be enhanced by adding *low-e* (low-emissivity spectrally selective) coatings—a common feature of modern glazing systems that adds additional insulating characteristics. Increasing the amount of GHGs in the atmosphere can be thought of as analogous to adding low-e coatings to window glazing. Improving the insulation in this way leads to an increase in the air temperature, in both the room and the analogous atmosphere. Essentially, that's all there is to the physics underpinning the GWH. It's relatively simple and is based on concepts that were proven experimentally in the nineteenth century.[11]

In addition to the Earth's radiation balance (the insulation of the greenhouse effect and the shading effects of clouds and aerosols), there are other factors that influence climate. One is how solar heating is divided between the oceans and the atmosphere. The oceans act as a giant heat sink, capable of absorbing a greater amount of heat than the atmosphere and releasing it slowly over time. The resulting thermal inertia, akin to concrete left out in the Sun, plays a major role in short-term climate fluctuations and slows the rate of climate change. Another factor is the spherical shape of the Earth. The Earth is heated more strongly at the equator than at the poles, resulting in vigorous circulations in the atmosphere and oceans that transmit heat poleward. These circulation patterns are affected by the rotation of the Earth and the shapes of the continents, resulting in the complex patterns of climate found around the globe. Fundamentally, however, the Earth's surface temperature is set by the radiative heat balance, i.e., energy in = energy out.

PART 1 OF THE GLOBAL WARMING HYPOTHESIS

The Earth's average temperature is rising due to an increase in the concentration of GHGs in the atmosphere resulting from human activity, principally the burning of fossil fuels and the clearing of forests.

Atmospheric GHG concentrations have been increasing since around 1750, the start of the European industrial revolution (Figure 1.4), having previously been stable over the last 10,000 years. With modern computer modeling methods, it has now been established that the increase in GHGs can explain the increase in global temperature.[12] The timing of the increase in GHGs suggests an anthropogenic origin but could have been a coincidence. Subsequent research into the Earth's carbon cycle (see "The Carbon Cycle" box) has shown that only human activity can explain the increase in CO_2.[13] Although it is more difficult to prove conclusively that the increased concentrations of some *non-CO2* GHGs have been entirely anthropogenic—the sources and chemical processes involved are complex—the evidence suggests this is very likely to be the case. These findings led the world's climate science community to issue, in 2007, the strongest and most definitive statement on the first part of the GWH that had been issued up to that date: "Most of the observed increase in global average temperatures since the mid-20th century is very likely [at least a 90 percent likelihood] due to the observed increase in anthropogenic GHG concentrations."[14]

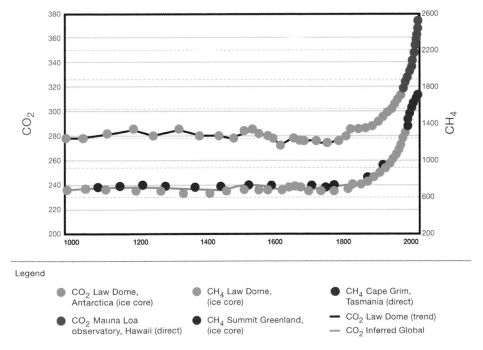

1.4 Variations in atmospheric CO_2 and CH_4 since the start of the industrial revolution; the data for the last few decades are from direct measurements, the remainder from analysis of Greenland and Antarctic ice cores—because CO_2 and CH_4 are well-mixed and long-lived gases in the atmosphere, point measurements give a good indication of global average values
Graphic: Arup, Data: NOAA Paleoclimatology (available online: http://www.ncdc.noaa.gov/paleo/data.html)

The Carbon Cycle

The level of atmospheric CO_2 is determined by the carbon cycle (Figure 1.5).[15] There are four main stores of carbon—the atmosphere (CO_2 in the air), the land biosphere (living organic matter and organic debris held in soil), the oceans (organic matter, calcium carbonate, and dissolved CO_2), and fossil fuel deposits—and four pathways through which carbon is continually transferred between these stores. Two are natural: absorption and release by the land biosphere (from photosynthesis and respiration) and release and absorption at the surface of the ocean (by photosynthesis and respiration of marine life, and chemical absorption and release in seawater). Two are entirely anthropogenic: deforestation and fossil fuel combustion. (A fifth process, outgassing from the mantle by volcanic activity, was very important during the Earth's past but has been insignificant in recent geological time.)

The carbon cycle resembles the Earth's thermal radiation budget: the level of CO_2 is determined by the flows in and out, and reaches equilibrium when they balance. Currently, humans add around 7 gigatons of carbon (GtC)[16] each year through burning fossil fuels and about 1.5 GtC from deforestation. If flows change, the carbon cycle eventually reaches a new equilibrium at a different level of atmospheric CO_2. This process is thought to take around 100 years. Because the carbon cycle has already been disturbed, stopping anthropogenic emissions of CO_2 today would not return atmospheric levels to preindustrial levels.

While the oceans and the land biosphere currently act as absorptive sinks for CO_2, slowing the rate of change, in the future they may become net sources of CO_2 for the atmosphere. The warming of the climate now underway will cause increased *natural* emissions, such as forest fires, decomposition of formerly frozen tundra, and outgassing of dissolved CO_2 from the ocean. If these increased emissions exceed the ability of the land and ocean to absorb CO_2, it will constitute a *positive feedback* for even faster warming.

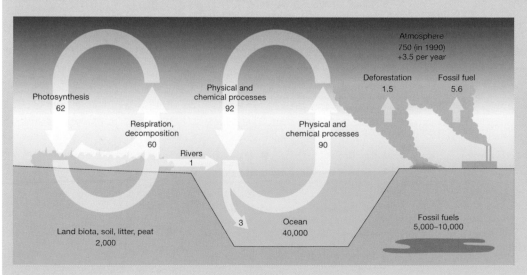

1.5 The carbon cycle: exchange of carbon with the atmosphere (as CO_2), on the land (in the bodies of plants and animals and in organic matter in soil and peat), in the ocean (as dissolved CO_2 and calcium carbonate), and in fossil fuel (reserves); fluxes are in gigatons of carbon (GtC) per year, stores in GtC
Graphic: Arup, adapted from J. Houghton, *Global Warming: The complete briefings*, 3rd ed. (Cambridge: Cambridge University Press, 2004)

HOW CERTAIN IS THE SCIENCE?

Despite the above conclusions, the concept of anthropogenic global warming is frequently portrayed in the popular media as something around which there is still much uncertainty and debate. From where does this skepticism originate?

Paradoxically, one source may have been the climate science community itself. The scientific community expresses its conclusions to a level of certainty that is not absolute, and this is often interpreted by nonscientists as leaving more room for doubt than really exists. The wording of the concluding statements of the Intergovernmental Panel on Climate Change (IPCC) is also shaped through international consensus building. In this light, the IPCC conclusion that Part 1 of the GWH is proved with at least 90 percent certainty should be considered more as a statement of fact than one of doubt.

In the lay community, sources of skepticism are varied. It can be difficult for the nonscientist to form a judgment because of the complexity of information. Some projections for global warming are truly frightening. It can be seductive to deny the evidence for change. Some of the arguments used to refute the GWH include the following:

1. **There is no global warming trend.**
   This is called *trend skepticism*. The main method here has been to find particular locations around the world where temperatures are not rising or may even be decreasing. It is important to realize, however, that temperature data from any given location contains "noise," particularly interannual and decadal climate variability. This is why global and regional averages are usually discussed when considering climate change. The averaging process reduces the noise of local climate variability, allowing the underlying trends to be revealed.

2. **The global warming trend is real but not due to the increase in GHGs.**
   This is *attribution skepticism*. The main method here is to propose another mechanism for the observed warming. The primary candidates are shading effects such as volcanic eruptions and cloudiness, and changes in heat input from variations in solar output. Volcanic eruptions tend to produce cooling episodes and do not provide an explanation for sustained warming. Solar output

increased slightly over the industrial period, but the changes are too small to directly explain the observations of warming. One theory, which remains a topic of active research but hasn't been well established, is that variations in solar output affect the Earth's cloudiness.[17] Ironically, these alternative attributions would combine with the well-documented effects of GHGs and further increase global warming, making the impacts even worse than IPCC projections. It is only the effects that can counteract the increased greenhouse effect that will slow or reverse the anthropogenic global warming. The only known effect of this type is actually anthropogenic in origin—the tendency of aerosol pollution from fossil fuel and biomass burning to provide a shading effect (so-called *global dimming*).

   Another line of attack is to suggest the climate models are fundamentally flawed in some way. However, as we've described above, although the details of rates of change and regional variations require sophisticated computer modeling, the physics central to the concept of GHG-driven climate change is simple and can essentially be calculated (admittedly with some effort) by hand. The magnitude of the estimates for the extent of warming that can be expected from changes in GHGs have remained remarkably robust over several decades.

3. **Global warming is just part of natural variability, and therefore would have happened anyway.**
   Another attribution skeptic approach is to show that climate can alter naturally without human intervention, so there is no need to assign blame to ourselves. While it is true that the Earth's climate has changed in the past due to natural variations in GHGs and in response to external factors, such as solar output changes, volcanic eruptions, and asteroid impact, all the evidence points to the current period of climate change being anthropogenic. Moreover, the observed rate of increase of greenhouse concentrations and the projected rate of global warming appear to be unprecedented in the geological record.

4. **Global warming will lead to a new ice age due to the Gulf Stream switching off.**
   This is a scenario popularized by the film *The Day After Tomorrow*. While there is evidence in the geological record that during ice ages global climate can flip quite quickly between warm and

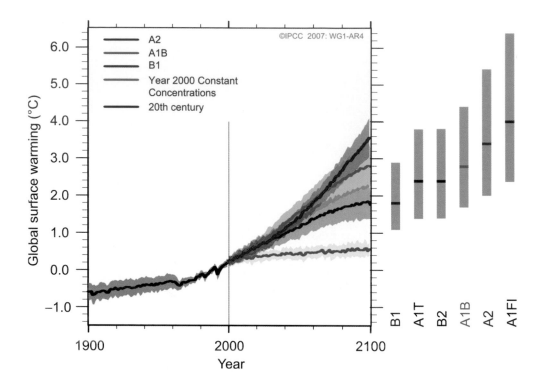

1.6 IPCC AR4 projections for global average near-surface temperature change under three of the SRES marker scenarios (A2, A1B, and B1) and indicative ranges for 2100 for all six SRES marker scenarios. Solid lines are multi-model global averages of surface warming (relative to the 1980–99 average) with shading denoting the ±1 standard deviation range; the orange line is for the experiment where concentrations were held constant at year 2000 values, and the gray bars at right indicate the best estimate (solid line within each bar) and the likely range assessed for all six SRES marker scenarios

Graphic: IPCC, Solomon et al., 2007[ii]

cool states due to changes in the ocean circulation in the North Atlantic,[18] it is likely that this instability in the climate system was governed by land and sea ice feedback processes. The relatively ice-free nature of the modern world makes this scenario much less probable today. Current climate models suggest a climate reversal is unlikely to happen under anthropogenic greenhouse forcing.[19]

5. **Global warming won't actually be that bad.**
   This is *impact skepticism*. It relates to the second part of the GWH and is covered below.

PART 2 OF THE GLOBAL WARMING HYPOTHESIS

If humans continue development based on fossil fuel energy, anthropogenic global warming and associated climate changes will accelerate, with dangerous consequences for both the human and natural world.

The first step in assessing this part of the GWH is to examine what the likely magnitude of climate change will be in the future. The second step is to examine what the impacts might be.

Climate Change Projections

Climate change models are complex computer models that have been developed to understand the possible course of future climate change and also to understand natural climate changes that have occurred in the past. These models aim to simulate the physical, chemical, and biological processes that affect the climate system, as far as computational restrictions allow. They are used by a number of climate modeling centers around the world.

The first step in the generation of climate change projections is the development of *emissions scenarios*. These are storylines for future emissions of GHGs and other anthropogenic factors affecting climate such as aerosol pollution. A widely used set of emissions scenarios are those in the IPCC Special Report on Emissions Scenarios (SRES).[20] The SRES scenarios use six storylines based on different assumptions regarding future demographic, economic, and technological trends. For each storyline, the SRES considered a range of possibilities, and a single illustrative or *marker* emissions scenario was provided for each, resulting in a set of six key scenarios (Table 1.3). Four of these storylines (the A group) assume a world where economic growth is the dominant driver, in a way similar to today. The remaining two (the B group) assume that the key driver is sustainability (economic, social,

TABLE 1.3

The six SRES marker emissions scenarios

Data: Nakićenović and Swart, 2000, see note 20

Scenario	Key features	Underlying assumptions	Atmospheric CO_2 concentration at 2100 (ppm)
A1F1	Economic growth Convergent world Fossil fuel energy	A convergent world with substantial reduction in regional differences in per capita income and rapid economic growth; a global population that peaks in mid-century and declines thereafter; rapid introduction of new technologies; energy sources based primarily on continuing use of fossil fuels	970
A2	Economic growth Heterogeneous world	A heterogeneous world with self-reliance and preservation of local identities; fertility patterns across regions converge very slowly, resulting in continuously increasing global population; economic development is primarily regionally oriented; per capita economic growth and technological change are more fragmented and slower than in other storylines	856
A1B	Economic growth Convergent world Mixed energy sources	As A1F1 but with energy based on a balance across fossil and non-fossil sources	717
B2	Sustainability Heterogeneous world	A heterogeneous world with local and regional solutions to economic, social, and environmental sustainability and social equity; continuously increasing global population but at a rate lower than A2; intermediate levels of economic development; less rapid and more diverse technological change than in the B1 and A1 storylines	621
A1T	Economic growth Convergent world Non-fossil fuel energy	As A1F1 but with energy based on non-fossil fuel sources	582
B1	Sustainability Convergent world	A convergent world with global solutions to economic, social, and environmental sustainability with an emphasis on improved equity; the same global population as in the A1 storyline, but with rapid changes in economic structures toward a service and information economy; reductions in material use intensity; introduction of clean and resource-efficient technologies.	549

and environmental). By 2100, the increased atmospheric CO_2 concentration in the scenarios ranges from two times (B1, 549 parts per million [ppm]) to three and a half times (A1F1, 970 ppm) preindustrial levels (280 ppm).

The SRES storylines do not assume the implementation of specific policies to stabilize atmospheric GHG concentrations for climate change mitigation. However, implicit in three of the scenarios (A1T, B1, and B2) is a move to lower-carbon fuels and greater energy efficiency, resulting in reduced per capita carbon emissions.

The six SRES marker scenarios have been widely used as the basis for climate change modeling studies. Figure 1.6 shows the projections from a range of climate models for changes in global near-surface air temperature to year 2100 under the SRES scenarios and also for a "2000 constant concentration" scenario, which is indicative of how the climate system would continue to adjust if

all the anthropogenic emissions of GHGs had stopped in the year 2000.[21] Estimated ranges of uncertainty are also shown. A feature of the graph is that the best-estimate projections for the different scenarios remain relatively close until around the middle of the century and then start to fan out. While the best-estimate values differ consistently between the scenarios, in order of increasing CO_2 concentrations, there is a substantial overlap in the uncertainty ranges for each. This indicates that the modeling uncertainty can be significant as well as the emissions-scenario uncertainty. The complete range of projected global temperature change, including the uncertainty ranges, is 1.1 to 6.4°C (2 to 11°F). It is important to realize that these changes in global temperature are very considerable. They are on par with some of the largest climate changes the Earth has seen in its history—moving between ice age periods, temperate periods (like today), and very warm periods (such as the dinosaur era).

Impacts

Assessing what the impact of climate change might be is even more difficult than assessing what the magnitude of the climate changes might be. We are dealing with things that may have never happened before, and if they did, they happened far in the past and left only sketchy evidence in the geological record. Whether or not the impacts will be "damaging" is a subjective assessment closely linked to the risk and coping capacity of those people and ecological systems impacted. On the scale of millions of years, it is reasonable to assume that the catastrophic damage climate change is likely to inflict in the next few hundred years will be all but forgotten as life emerges in a different, adapted form. However, for all of us with families, friends, and dreams of a human future, *damaging* seems an appropriate adjective for what science suggests lies ahead.

Perhaps the best way to understand the projected impacts is actually to put them in the context of the recent geological past transposed on the soon-to-be future. Take, for example, an impact event where a meteor strikes the Earth. We'll schedule it for January 2052, roughly 40 years from now, when most of those we know are still around to experience it.

Within 24 to 72 hours of the impact, sea levels around the world have risen by nearly 3 meters (10 feet), flooding coastal cities and villages.

In the weeks that follow, debris and gases are released into the air, causing average temperatures to climb by 2 to 5°C (4 to 9°F). Rainfall and elevated temperatures in mountainous regions accelerate the melting of snowpack, resulting in severe flooding along the world's rivers and depletion of freshwater storage for drinking, agriculture, and industry. City levees and infrastructure are overwhelmed.

In the months that follow, temperatures continue to rise. Hurricanes and typhoons encouraged by the warm waters and turbulent atmosphere increase in ferocity, sending waves and winds against coastal communities and offshore oil platforms. Precipitation declines in inland areas, causing widespread water shortages, crop failure, and blinding dust storms.

In the years that follow, ocean acidification has largely killed off and started to dissolve coral reefs, leading to a collapse of the marine food chain. On land, insects have taken advantage of weak plants and warm winters, resulting in deep losses in forestry, farmland, and buildings. Lightning ignites wildfires, which are spread by winds and fed by excess dead biomass. Millions of people have been dislocated. Over 50 percent of species have gone extinct. Lowered quality of air, water, and food contributes to an increase in disease. Trillions of dollars of damage has destabilized the world financial systems, resulting in mass unemployment and depression.

The only real difference between an impact event and the GWH is the timescale. Rather than taking place in the span of five years, the impacts will distribute over many decades. That timescale may actually make all the difference. Although the long timescales of the GWH won't hurry people to action in quite the same way that an anticipated meteor strike would, there is a positive side. The additional time provides an opportunity to reduce the severity of the impacts and prepare for their coming. There is a need for urgency, however, because the impacts of the additional GHGs that have already been added to the atmosphere, and those that will be added over the coming decades, will persist for many decades and centuries to come, while the climate system adjusts to reach a new equilibrium. Whether we choose to meet this challenge is for our generation to decide. It will be necessary to both reduce the emissions of GHGs (*mitigation*) and manage the impacts of unavoidable climate changes (*adaptation*). A path to succeed at both is described in Parts 2 and 3 of this book.

NEW KNOWLEDGE SINCE 2006

Consensus takes time. This is especially true of the extensive peer review and fact-checking required of scientific documents, like those of the IPCC. Indeed, the highly referenced IPCC *Fourth Assessment Report* (AR4) in 2007 considered only work published before July 2006.[22] And the IPCC AR5 report is not due for publication until 2013. Fortunately, climate research, like the changing climate itself, is not static. So, what's happened since 2006? There have been hundreds of interim research publications. Despite local weather variability and a decrease in solar forcing over the last decade, a combination of scientifically documented changes and improved modeling techniques since 2006 suggests that climate change is accelerating and human-induced emissions are the cause. The 2007

IPCC report projections increasingly appear to reflect a conservative underestimate.

This new knowledge serves as a check of our collective responsiveness to earlier reports and is an opportunity for new insights. What follows is a summary of the new knowledge since the last IPCC report. It is a synopsis of reviews by the Center for Climate and Energy Solutions (formerly the Pew Center on Global Climate Change), the 2009 *Copenhagen Diagnosis*, and other later sources.[23]

- **Surging GHG emissions**
  Through 2007, the growth rate of actual CO_2 emissions has tracked the most pessimistic (i.e., the fastest rate) of the IPCC scenarios.[24] In 2008, global CO_2 emissions from fossil fuels were 40 percent higher than those in 1990. Then in 2010, CO_2 emissions jumped by the biggest amount on record, resulting in levels of GHGs higher than the worst-case scenario published by the IPCC.[25]
- **Predictions for increased future sea-level rise**
  By 2100, global sea level is likely to rise at least twice as much as projected by Working Group I of the IPCC AR4. The most current upper limit has been estimated as 2 meters (7 feet) by 2100. Sea level will continue to rise for centuries after global temperatures have been stabilized, and many meters (more than 10 feet) of sea-level rise should be expected over the next few centuries.[26]
- **Non-uniform sea-level rise**
  Two studies found that sea level does not rise uniformly around the world. For example, the Pacific and Atlantic coasts of the United States will experience significantly more sea-level rise than the global average.[27]
- **Catastrophic abrupt sea-level rise is more likely than previously understood**
  It has been shown that collapse of the West Antarctic Ice Sheet will cause global sea levels to rise rapidly by more than 3 meters (10 feet).[28]
- **Doubling of average global temperature increase by 2100**
  Updated models show twice as much global warming at the end of the current century: 5.2°C (9.4°F) in the updated model run compared to 2.4°C (4.3°F) before the update.[29]

- **Accelerating Arctic sea ice decline**
  Summertime melting of Arctic sea ice has accelerated beyond the predictions of climate models. The area of summertime sea ice during 2007 to 2009 was about 40 percent less than the average prediction from IPCC AR4 climate models.[30]
- **Observed ecosystem changes increasing**
  Climate change is already affecting multiple systems, both physical (e.g., timing of seasonal lake freeze and thaw) and biological (e.g., timing of seasonal plant flowering and animal migration).[31] Global precipitation trends linked to human emissions are larger than model predictions and are likely already affecting economy, agriculture, and human health in certain regions.[32]
- **Ocean acidification damage to fisheries, tourism, and coastal protection better understood**
  Acidification caused by CO_2 gas dissolving to form carbonic acid "may render most regions chemically inhospitable to coral reefs" by 2050.[33] Since many fisheries depend on corals and shell-forming organisms to support their food chains, ocean acidification is a threat to national food security, tourism, and coastal protection.[34] A recent modeling study concluded that by the time atmospheric concentrations of CO_2 reach 560 ppm, "all coral reefs will cease to grow and start to dissolve."[35]
- **Runaway warming is more likely but still poorly understood**
  Accelerated melting of ice[36] and permafrost[37] suggest increased natural forcing for global warming, specifically as a result of reduced surface reflectivity from ice and greater methane release from decomposition.
- **Changes more likely to be irreversible for over 1,000 years**
  New studies find that warmer temperatures and changes in precipitation caused by CO_2 emissions from human activity are largely irreversible.[38]

AVOIDING DANGEROUS CLIMATE CHANGE

For some time, it has been considered that a change in global temperatures of more than 2°C (3.6°F) compared to preindustrial levels will put us in the realm of dangerous and unavoidable climate change. As can be seen from Figure 1.6, this level is projected to be exceeded by all of the "best guess" IPCC projections under the SRES emissions scenarios. At what level, then, would GHG concentrations need to stabilize to avoid this threshold, and how quickly would emissions need to come down to achieve these levels?

It is now generally believed that we must stabilize GHGs at 450 ppm CO_2 equivalent (CO_2e) or below in order to avoid 2°C climate change, and even then there is a low but significant probability the threshold will be exceeded.[39] To achieve 450 ppm CO_2e requires that global emissions start reducing immediately and reach well under 1 metric ton CO_2 per capita by 2050 (80 to 95 percent below per capita emissions in developed nations in 2000). Further, it is now believed that the previous 2°C target may be too high, more representative of a threshold between dangerous and extremely dangerous climate change, since the impacts of climate change are now expected to be larger than previously thought.[40] Some authors have argued that a stabilization limit of 350 ppm CO_2e is necessary to avoid dangerous climate change.[41] Since this is *below* the current level of GHGs in the atmosphere, some active action to *remove* (sequester) CO_2 from the atmosphere would be required in addition to emission reductions to achieve this level. Based on current trends, the chances of achieving stabilization anywhere below about 550 ppm are looking remote unless very aggressive action is taken.

THE ECONOMIC AND SOCIAL CONSEQUENCES OF INACTION FOR THE BUILT ENVIRONMENT

In late 2006, the most complete work to date on the global economic impacts of climate change was published by Nicholas Stern, chair of the Grantham Research Institute on Climate Change and the Environment at the London School of Economics. Although the report stirred debate among economists, its quality has largely been upheld. Among the Stern Review conclusions was an assertion that "the benefits of strong, early action on climate change far outweigh the costs of not acting."[42] The costs of inaction calculated by the Stern Review team ranged from an optimistic 5 percent of global gross domestic product (GDP) to a pessimistic 20 percent. In comparison, the costs of action were calculated to be approximately 2 percent of global GDP.[43]

Most scenarios show that developing nations face disproportionate negative impacts from climate change due largely to their reduced coping capacity (see Chapter 16 for further discussion). Although risks to life in developed countries are less due to greater coping capacity, the greater value of building assets is a clear and present risk that is likely to stimulate local action.

Although there has been historic resistance to climate change legislation due to concerns over national economic health and taxation, recent years have seen a significant reversal in attitude toward revenue-neutral taxation and green-collar investment. A report by the United Nations in September 2008 concluded that "investments in improved energy efficiency in buildings could generate an additional 2–3.5 million green jobs in Europe and the United States alone, with the potential much higher in developing countries." The report further stated that the global market for environmental products and services is projected to double from U.S.\$2.74 trillion by 2020, with half of the growth in efficiency and the balance in sustainable transport, water supply, sanitation, and waste management.[44] Additional research sponsored by the U.S. Conference of Mayors in 2008 forecasts that, by 2038, the economy will generate 4.2 million new green jobs, making up 10 percent of new job growth.[45]

THE SCIENCE OF CLIMATE CHANGE IN SUMMARY

This chapter has outlined the scientific basis for the GWH and presented the main scientific evidence to support its two parts—(1) the attribution of contemporary global warming to anthropogenic GHGs, and (2) the likelihood of significant damage and irreversible impacts of accelerating emissions on humans and the Earth's ecology. The evidence for both is now robust. The increase in GHGs is anthropogenic and provides an entirely credible explanation for the observed warming trends; no other known processes provide an alternative explanation, and no known process is overriding the GHG warming effect or is likely to in the near future. The corollary is that if human beings continue to produce GHGs in large quantities, the warming trend will continue, with impacts that should be considered severe enough to justify immediate action and investment. Climate change is a very real and immediate danger.

NOTES

1 The average land and ocean near-surface air temperature.

2 S. Solomon, D. Qin, M. Manning, Z. Chen, M. Marquis, K. B. Averyt, M. Tignor, and H. L. Miller, eds., *Contribution of Working Group I to the Fourth Assessment Report of the Intergovernmental Panel on Climate Change, 2007* (New York: Cambridge University Press, 2007).

3 NASA, "GISS Surface Temperature Analysis," http://data.giss.nasa.gov/gistemp/ (accessed June 4, 2012); Met Office, "HadCRUT3 dataset," http://www.metoffice.gov.uk/hadobs/hadcrut3/ (accessed June 4, 2012).

4 See note 2 above; I. Allison, N. L. Bindoff, R. A. Bindschadler, P. M. Cox, N. de Noblet, M. H. England, J. E. Francis, et al., *The Copenhagen Diagnosis, 2009: Updating the World on the Latest Climate Science* (Sydney, Australia: The University of New South Wales Climate Change Research Centre, 2009), http://www.copenhagendiagnosis.com/ (accessed June 4, 2012).

5 Joint Science Academies (G8, China, India, Brazil), *Joint Science Academies' Statement: Global Response to Climate Change*, 2005, http://www.nationalacademies.org/onpi/06072005.pdf (accessed June 4, 2012); U.S. Global Change Research Program, *Global Climate Change Impacts in the United States* (New York: Cambridge University Press, 2009), http://downloads.globalchange.gov/usimpacts/pdfs/climate-impacts-report.pdf (accessed June 4, 2012).

6 A scientific and working hypothesis is a proposed explanation for a phenomenon that is testable and provisionally accepted for further research. The GWH meets these criteria.

7 Spencer Weart, *The Discovery of Global Warming*, 2nd ed., 2003, http://www.aip.org/history/climate/index.html (accessed June 4, 2012).

8 Arrhenius expected CO_2 doubling to take about 3,000 years, based on the rate of coal use at the time; it is now estimated in most scenarios that a doubling of atmospheric CO_2 concentrations will occur at some point this century.

9 Roger Revelle and Hans E. Suess, "Carbon Dioxide Exchange between Atmosphere and Ocean and the Question of an Increase of Atmospheric CO_2 During the Past Decades," *Tellus* 9(1), 1957: 18–27.

10 National Academy of Sciences, "Carbon Dioxide and Climate: A Scientific Assessment" report to the Climate Research Board, Assembly of Mathematical and Physical Sciences, National Research Council (Washington, DC: National Academy of Sciences, 1979), http://books.nap.edu/catalog.php?record_id=12181 (accessed June 25, 2012).

11 See note 7 above.

12 See note 2 above.

13 J. T. Houghton, Y. Ding, D. J. Griggs, M. Noguer, P. J. van der Linden, X. Dai, K. Maskell, et al., eds., *Climate Change 2001: The Scientific Basis, Contribution of Working Group I to the Third Assessment Report of the Intergovernmental Panel on Climate Change* (New York: Cambridge University Press, 2001).

14 See note 2 above.

15 C. B. Field and M. Raupach, eds., *The Global Carbon Cycle: Integrating Humans, Climate, and the Natural World* (Washington, DC: Island Press, 2004).

16 The weight of carbon is used when discussing the carbon cycle because a common unit is needed that relates its different forms in the different stores (carbon is in the form of CO_2 only in the atmosphere, generally). To convert from weight of CO_2 to weight of carbon, one needs to divide by $^{44}/_{12} = 3.7$, the ratio of the molecular weight of a CO_2 molecule to that of a carbon atom (i.e., 1 ton carbon = 3.7 tons CO_2).

17 See note 2 above.

18 Richard B. Alley, *The Two-Mile Time Machine: Ice-Cores, Abrupt Climate Change and Our Future* (Princeton, NJ: Princeton University Press, 2000).

19 See note 2 above.

20 Nebojsa Nakićenović and Rob Swart, eds., *Emissions Scenarios*, special report, IPCC (New York: Cambridge University Press, 2000).

21 See note 2 above.

22 See note 2 above.

23 Center for Climate and Energy Solutions (formerly the Pew Center on Global Climate Change), *Key Scientific Developments Since the Fourth IPCC Assessment Report*, 2009, http://www.c2es.org/docUploads/Key-Scientific-Developments-Since-IPCC-4th-Assessment.pdf (accessed June 5, 2012); and see note 4 above (Allison et al., 2009). The *Copenhagen Diagnosis* was written by twenty-six climate scientists, all active researchers, from eight countries. The group of authors is independent and unaffiliated with any organization. About half are IPCC authors, with firsthand experience in preparing such an assessment and an understanding of the scientific standards it should meet. The report is firmly based on the more than two hundred cited peer-reviewed papers.

24 J. G. Canadell, C. Le Quéré, M. R. Raupach, C. B. Field, E. T. Buitehuis, P. Ciais, T. J. Conway, N. P. Gillett, R. A. Houghton, and G. Marland,

"Contributions to Accelerating Atmospheric CO_2 Growth from Economic Activity, Carbon Intensity, and Efficiency of Natural Sinks" *PNAS* 104(47), 2007: 18,866–70.

25 See Carbon Dioxide Information Analysis Center, http://cdiac.ornl.gov/.

26 The IPCC sea-level rise projections of 0.2 to 0.6 meters (0.6 to 1.9 feet) by 2100 were based largely on thermal expansion models of the world's oceans and did "not assess the likelihood, nor provide a best estimate or an upper bound for sea-level rise" (R. K. Pachauri and A. Reisinger, *Contribution of Working Groups I, II and III to the Fourth Assessment Report of the Intergovernmental Panel on Climate Change* [Geneva, Switzerland: IPCC, 2007]). The upper end of the IPCC projection should not, therefore, be interpreted as an upper limit; in fact, recent evidence finds that the IPCC *under*estimates future sea-level rise (S. Rahmstorf, "A Semi-Empirical Approach to Projecting Future Sea-Level Rise," *Science* 315(5810), 2007: 368–70; S. Jevrejeva, J. C. Moore, A. Grinsted, and P. L. Woodworth, "Recent Global Sea Level Acceleration Started Over 200 Years Ago?" *Geophysical Research Letters* 35(8), 2008: 8715).

27 J. L. Bamber, R. E. M. Riva, B. L. A. Vermeersen, and A. M. LeBrocq, "Reassessment of the Potential Sea-Level Rise from a Collapse of the West Antarctic Ice Sheet," *Science* 324(5929), 2009: 901–3; J. Yin, M. E. Schlesinger, and R. J. Stouffer, "Model Projections of Rapid Sea-Level Rise on the Northeast Coast of the United States," *Nature Geoscience* 2(4), 2009: 262–6.

28 See note 27 above (Bamber et al., 2009).

29 Andrei P. Sokolov, P. H. Stone, C. E. Forest, R. Prinn, M. C. Sarofim, M. Webster, S. Paltsev, et al., "Probabilistic Forecast for 21st Century Climate Based on Uncertainties in Emissions (Without Policy) and Climate Parameters," *Journal of Climate* 22(19), 2009: 5175–204.

30 National Snow and Ice Data Center, "Antarctic Ice Shelf Disintegration Underscores a Warming World," March 25, 2008, http://nsidc.org/news/press/20080325_Wilkins.html (accessed June 5, 2012); National Snow and Ice Data Center, "Media Advisory: Ice Bridge Supporting Wilkins Ice Shelf Collapses," April 9, 2009, http://nsidc.org/news/press/20090408_Wilkins.html (accessed June 5, 2012); T. L. Mote, "Greenland Surface Melt Trends 1973–2007: Evidence of a Large Increase in 2007," *Geophysical Research Letters* 34, 2007: L22507; M. Tedesco, "A New Record in 2007 for Melting in Greenland," *Eos, Transactions American Geophysical Union* 88, 2007: 39; J. Stroeve, M. M. Holland, W. Meier, T. Scambos, and M. Serreze, "Arctic Sea Ice Decline: Faster Than Forecast," *Geophysical Research Letters* 34, 2007: 9501; J. Boé, A. Hall, and X. Qu, "September Sea-Ice Cover in the Arctic Ocean Projected to Vanish by 2100," *Nature Geoscience* 2, 2009: 341–3; M. Wang and J. E. Overland, "A Sea Ice Free Summer Arctic within 30 Years?" *Geophysical Research Letters* 36, 2009: L07502; E. J. Steig, D. P. Schneider, S. D. Rutherford, M. E. Mann, J. C. Comiso, and D. T. Shindell, "Warming of the Antarctic Ice-Sheet Surface Since the 1957 International Geophysical Year," *Nature* 457, 2009: 459–62; S. H. Mernild, G. E. Liston, C. A. Hiemstra, and K. Steffen, "Record 2007 Greenland Ice Sheet Surface Melt Extent and Runoff," *Eos, Transactions American Geophysical Union* 90, 2009: 13–14.

31 C. Rosenzweig, D. Karoly, M. Vicarelli, P. Neofotis, Q. Wu, G. Casassa, A. Menzel, et al., "Attributing Physical and Biological Impacts to Anthropogenic Climate Change," *Nature* 453, 2008: 353–7.

32 X. Zhang, F. W. Zwiers, G. C. Hegerl, F. H. Lambert, N. P. Gillett, S. Solomon, P. A. Stott, et al., "Detection of Human Influence on Twentieth-Century Precipitation Trends," *Nature* 448, 2007: 461.

33 "Monaco Declaration: Second International Symposium on the Ocean in a High-CO_2 World," 2008, http://www.ocean-acidification.net/Symposium2008/MonacoDeclaration.pdf (accessed June 25, 2012).

34 O. Hoegh-Guldberg, P. J. Mumby, A. J. Hooten, R. S. Steneck, P. Greenfield, E. Gomez, C. D. Harvell, et al., "Coral Reefs Under Rapid Climate Change and Ocean Acidification," *Science* 318, 2007: 1737.

35 J. Silverman, B. Lazar, L. Cao, K. Caldeira, and J. Erez, "Coral Reefs May Start Dissolving When Atmospheric CO_2 Doubles," *Geophysical Research Letters* 36, 2009: L05606.

36 See note 30 above.

37 K. M. Walter, L. C. Smith, and F. S. Chapin III, "Methane Bubbling from Northern Lakes: Present and Future Contributions to the Global Methane Budget," *Philosophical Transactions of the Royal Society A: Mathematical, Physical and Engineering Sciences* 365, 2008: 1657–76; C.-L. Ping, G. J. Michaelson, M. T. Jorgenson, J. M. Kimble, H. Epstein, V. E. Romanovsky, and D. A. Walker, "High Stocks of Soil Organic Carbon in the North American Arctic Region," *Nature Geoscience* 1, 2008: 615–19; E. A. G. Schuur, J. Bockheim, J. G. Canadell, E. Euskirchen, C. B. Field, S. V. Goryachkin, S. Hagemann, et al., "Vulnerability of Permafrost Carbon to Climate Change: Implications for the Global Carbon Cycle," *BioScience* 58, 2008: 701–14.

38 H. D. Matthews and K. Caldeira, "Stabilizing Climate Requires Near-Zero Emissions," *Geophysical Research Letters* 35, 2008: L04705; S. Solomon,

G. K. Plattner, R. Knutti, and P. Friedlingstein, "Irreversible Climate Change Due to Carbon Dioxide Emissions," *Proceedings of the National Academy of Sciences* 106, 2009: 1704–9; M. Eby, K. Zickfeld, A. Montenegro, D. Archer, K. J. Meissner, and A. J. Weaver, "Lifetime of Anthropogenic Climate Change: Millennial Time Scales of Potential CO_2 and Surface Temperature Perturbations," *Journal of Climate* 22, 2009: 2501–11.

39 K. Andereson and K. Bows, "Beyond 'Dangerous' Climate Change: Emission Scenarios for a New World," *Philosophical Transactions of the Royal Society A: Mathematical, Physical and Engineering Sciences* 369(1934), 2011: 20–44.

40 See note 4 above (Allison et al., 2009).

41 James Hansen, Makiko Sato, Pushker Kharecha, David Beerling, Robert Berner, Valerie Masson-Delmotte, Mark Pagani, et al., "Target Atmospheric CO_2: Where Should Humanity Aim?" *The Open Atmospheric Science Journal* 2(1), 2008: 217–31.

42 *Stern Review: The Economics of Climate Change*, 2006, http://webarchive. nationalarchives.gov.uk/+/http://www.hm-treasury.gov.uk/independent_ reviews/stern_review_economics_climate_change/stern_review_report. cfm (accessed June 5, 2012).

43 Stern revised the earlier estimate of 1 percent of GDP to 2 percent of GDP in 2008.

44 United Nations Environment Programme, *Green Jobs: Towards Decent Work in a Sustainable, Low-Carbon World* (Washington, DC: Worldwatch Institute, 2008).

45 *US Metro Economies: Current and Potential Green Jobs in the US Economy*, prepared for the United States Conference of Mayors and the Mayors Climate Protection Center, 2008.

i   J. T. Kiehl and K. E. Trenberth, "Earth's Annual Global Mean Energy Budget," *Bulletin of the American Meteorological Society*, 78, 1997: 197–208.

ii S. Solomon, D. Qin, M. Manning, Z. Chen, M. Marquis, K. B. Averyt, M. Tignor, and H. L. Miller, eds., *Contribution of Working Group I to the Fourth Assessment Report of the Intergovernmental Panel on Climate Change*, 2007 (New York: Cambridge University Press).

100,000 participants and marchers joined the UN Climate Change Conference Demonstration from the Danish parliament to Bella Center, on December 12, 2009 in Copenhagen, Denmark

Image: Piotr Wawrzyniuk/Shutterstock.com

Greenhouse Gas Emissions in the Built Environment

Fiona Cousins

We the undersigned, senior members of the world's scientific community, hereby warn all humanity of what lies ahead. A great change in our stewardship of the earth and the life on it is required, if vast human misery is to be avoided and our global home on this planet is not to be irretrievably mutilated.

"World Scientists' Warning to Humanity," 1992[1]

All countries have to report on their carbon emissions as part of the United Nations Framework Convention on Climate Change (UNFCCC)—an agreement made at the Earth Summit in Rio in 1992. The parties to this treaty meet annually at the Conference of Parties (COP). The Kyoto Protocol, which seeks to reduce greenhouse gas (GHG) emissions, was negotiated at COP3 in 1997. Under the UNFCCC, Annex I Countries (generally developed countries) have to produce an annual inventory for each GHG showing the amount emitted and its source. Other countries (including China and less developed countries) have to produce a national communication that provides similar information but with less detail and less frequently.

The percentages of global anthropogenic GHG emissions from 2004 are shown in Figure 2.1, in terms of CO_2 equivalent (CO_2e). This chapter discusses each of these groups of gases in turn, using the *Inventory of US Greenhouse Gas Emissions and Sinks: 1990–2009*[2] to show the effect of the built environment as a whole on GHG emissions. Results for other industrialized countries will be approximately similar, and significant differences between countries have been pointed out in the text.

There are only a few activities that contribute to net GHG emissions, by acting as either a source or a sink. The gases of concern are carbon dioxide (CO_2) itself, methane (CH_4), nitrous oxide (N_2O), and fluorinated gases (F-gases). The activities responsible for the generation or absorption of CO_2 are divided into seven major categories in the Intergovernmental Panel on Climate Change's (IPCC) GHG inventory: electricity generation; industry; commercial buildings; residential buildings; agriculture; transportation (all of which act only as sources); and land use, land-use change, and forestry (LULUCF), which can act as either a source or a sink. As electricity generation is required only because of the activities in the other

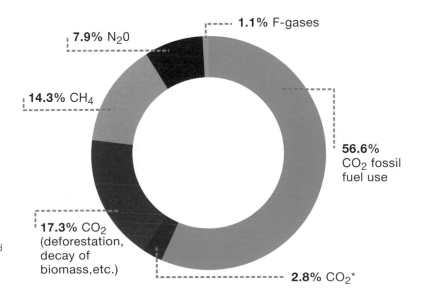

2.1 Global anthropogenic greenhouse gas emissions, 2004, in terms of CO_2
*CO_2 category including gas-well flaring and cement production
Graphic: Arup, Data: IPCC, 2007

7.9% N_2O

1.1% F-gases

14.3% CH_4

56.6% CO_2 fossil fuel use

17.3% CO_2 (deforestation, decay of biomass,etc.)

2.8% CO_2*

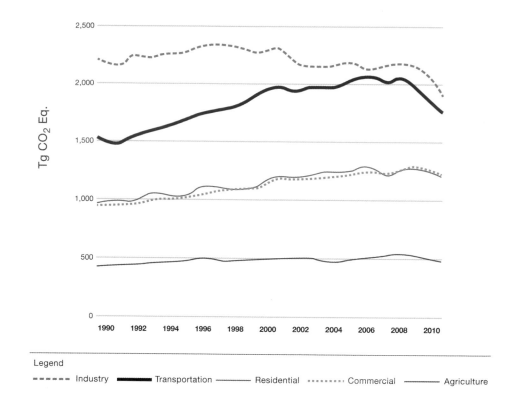

2.2 Inventory of U.S. greenhouse gas emissions and sinks, 1990–2009
Graphic: Arup, Data: EPA

Legend

------ Industry ▬▬▬ Transportation —— Residential ········· Commercial —— Agriculture

sectors, the emissions from electricity generation are then usually apportioned between the other five source categories. Figure 2.2 shows the relative emissions for the different sectors for the United States, with electricity apportioned to other categories.

This pattern, with industry as the most significant single source, followed by transportation, residential and commercial buildings, and then agriculture, is reflected in almost every country; in a few nonindustrialized nations, agriculture is the primary emissions source. If the residential and commercial sectors are combined, then buildings become the most significant source of emissions.

Buildings are not the only component of the built environment—the roads, railways, and pathways that connect buildings together and to their supply of goods compose the built environment as well. It also includes public parks and courts, private gardens, and other constructed open spaces that make towns and cities function. All of these elements are subject to design by planners, architects, and engineers. There are also components of industrial and transportation carbon emissions that are strongly related to the built environment and can be influenced by planning and design. If these emissions are also added to a built environment category, then this category is responsible for well over half of the United States' carbon emissions.

COMBUSTION-RELATED CO_2 EMISSIONS

The residential and commercial carbon emissions shown in the carbon inventory in Figure 2.2 include the emissions related to the fuel burned within buildings to produce heat or for cooking, and the emissions related to the electricity used within buildings. The transportation carbon emissions include the emissions related to fuel burned to make vehicles move and the emissions related to electricity used for transportation, for example in electric trains or buses. Together, these two groups of emissions make up 56.6 percent of the CO_2e emissions, as shown in Figure 2.1.

In all cases, the emissions are due to burning fossil fuels. This results in CO_2 emissions and N_2O emissions. The CO_2 emissions arise from the chemical reaction that takes place in combustion and are unavoidable. For natural gas, the reaction is as follows:

$$CH_4 \ + \ 2O_2 \ \rightarrow \ CO_2 \ + \ 2H_2O$$
methane   oxygen   carbon dioxide   water

The N_2O emissions[3] are generally a small proportion of the total emissions and are due to fuel impurities and to heat within the burner causing nitrogen in the air to oxidize. These are also unavoidable—they depend on both the fuel and the way in which the fuel is burned. Much of the regulation of emissions over the last several years, by the Environmental Protection Agency in the United States and by other agencies in other countries, is

TABLE 2.1
Emission levels for combustion of natural gas, oil, and coal
Data: U.S. Energy Information Administration, 1998[4]

Pollutant	Emission levels (pounds per billion Btu of energy input)		
	Natural Gas	Oil	Coal
Carbon dioxide	117,000	164,000	208,000
Carbon monoxide	40	33	208
Nitrogen oxides	92	448	457
Sulfur dioxide	1	1,122	2,591
Particulates	7	84	2,744
Mercury	0.000	0.007	0.016

aimed at reducing sulfur dioxide and nitrogen oxides emissions to minimize acid rain and smog formation, but it has also had a small effect on reducing global warming. This regulation has led to the introduction of "low-NO_x burners" for boiler installations and electricity generators, and catalytic converters for vehicles, which reduce the emissions of unburned fuel, carbon monoxide, and nitrogen oxides.

Table 2.1 shows that natural gas (CH_4) is the cleanest-burning fuel; of the three major fossil fuels used in buildings and electricity generation, it results in the lowest emissions of chemicals other than CO_2. Table 2.1 also shows that CO_2 emissions from burning natural gas are lower than CO_2 emissions from burning coal for the same energy input. One of the main implications of this is that electricity produced from natural gas has a lower carbon emissions factor than electricity produced from coal, all other things being equal. Conversion of power generation from coal to gas is one of

2.3 Carbon emissions factor for UK electricity grid; decline is primarily due to change from coal to gas generators
Graphic: Arup, Data: UK Department of Energy and Climate Change

the ways in which some countries, including the United Kingdom and the United States, have reduced their overall carbon emissions in the last twenty years (see Figure 2.3).

Within the transportation sector in the United States, approximately 65 percent of the carbon emissions are due to gasoline for vehicles, with the remaining 35 percent associated with internal airplane travel and freight traffic.[5] For smaller countries with less internal airplane travel, more of the carbon emissions are associated with road transport. This is because international airplane travel is excluded from the GHG inventories for individual countries.[6] Figure 2.2 illustrates that emissions associated with travel in the United States have risen rapidly in the last 20 years, as a result of relatively cheap fuel, which makes travel affordable, coupled with a lack of significant improvements in fuel efficiency.[7] Other industrialized countries where fuel is more expensive and where engine efficiency has been improved, usually as a result of policy, show less of an increase. Developing nations, where cars are just becoming affordable, will have an increase, although the magnitude of this is hard to derive from their infrequent UNFCCC national communications.

The residential and commercial emissions shown in Figure 2.2 are due either to electricity use attributed to buildings or to direct combustion of fossil fuels for building heating, water heating, or cooking. Together these are approximately 40 percent of annual carbon emissions, with small variations between industrialized countries. Chapters 7 and 9 discuss the means by which these emissions can be reduced. Key measures to reduce carbon emissions, which are evenly split between residential and commercial buildings, include reducing the amount of energy used within the building for all purposes: heating, cooling, office machinery, lighting, and ventilation. There are also significant changes to be made by changing the electricity source, either at the power station itself, or through the addition of district systems or substitution of renewable power sources for fossil fuels. These are covered in Chapter 8.

Case Study:
One Building, Many Choices—Total Emissions Footprint

As part of an exhibition developed for the American Institute of Architects New York chapter, we developed the concept of a total emissions footprint to help visualize the relative importance of transportation, operational, and embodied energy use. To develop a set of realistic numbers, we took a large commercial building program, designed to meet current energy and construction standards, and calculated the changes to the total emissions footprint through seven sets of changes to the building. These figures are based on publicly available data and in-house calculations, and are obviously only indicative, but provide some key insights into the relative effects of different design decisions.

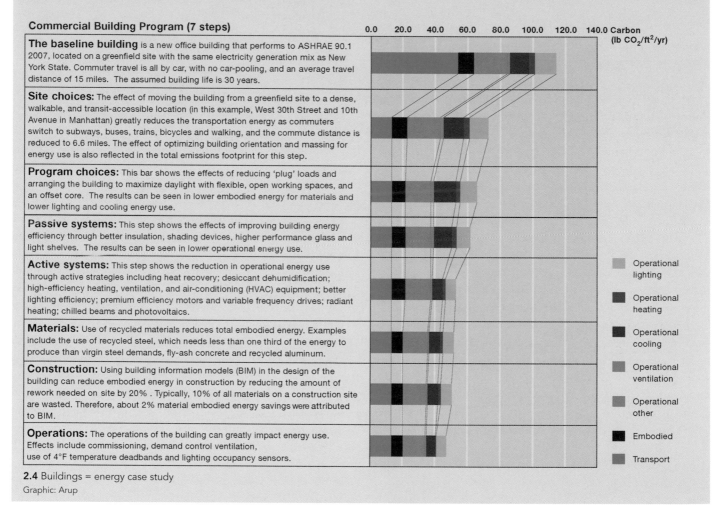

2.4 Buildings = energy case study
Graphic: Arup

OTHER CO$_2$ EMISSIONS

Design of the built environment can also have a significant influence on transportation energy, as shown in Figure 2.4. Details of the ways in which urban spaces can be designed to reduce emissions are discussed in Chapter 8—the key influence of built environment design is to reduce the vehicle miles traveled. This can be achieved through the adoption of transit-oriented development in which goods and services are provided via a pleasant walk or cycle from homes and workplaces.

Figure 2.1 shows two other segments related to CO$_2$ emissions, making up an additional 20.1 percent of global annual carbon emissions. The five largest contributors to these emissions in the United States are iron and steel production, flaring of gas wells, cement production, natural gas systems, and incineration of waste.

This category also includes LULUCF. Its driving force is the amount of carbon stored in biomass, which grows through photosynthesis, using CO$_2$ and water as its two major inputs:

Embodied Emissions

All construction materials have an embodied emissions footprint, which is the sum of emissions associated with the processes of extraction, manufacture, and transport. The total embodied energy of individual materials varies widely; however, there are just a few materials that make up the bulk of a building's embodied emissions, either because they are used extensively or because they have a high emissions footprint.

For example, aluminum is one of the most energy-intensive materials to produce—it requires over 33 times the embodied energy of hardwood per unit weight and over 7.5 times the energy input of recycled aluminum. Some comparative figures are provided in Figure 2.5, expressed in terms of the energy that would be produced by burning 1 U.S. gallon (3.8 liters) of oil. To reduce carbon emissions, choose building materials that are recycled, local, and minimally processed.

cubic feet	
0.01	virgin aluminum
0.02	virgin steel
0.13	glass
0.14	recycled steel
0.80	drywall
1.00	ceramic brick
1.15	concrete
3.33	hardwood
5.31	fiberglass insulation

1 gallon of oil is equal to …

2.5 Embodied emissions footprint of construction materials
Graphic: Arup, Data: various sources[ii]

$$6CO_2 + 6H_2O \rightarrow C_6H_{12}O_6 + 6O_2$$

carbon dioxide   water   sugars   oxygen

If the carbon is stored in biomass, it cannot be a component of CO_2. LULUCF can have a very variable impact on national emissions—for example, for the United States and Argentina, changes in land use to increase biomass reduce overall carbon emissions by about 15 percent; for Zimbabwe, LULUCF results in the entire country operating as a carbon sink; while for Indonesia, deforestation is the biggest element of the GHG inventory, taking carbon emissions from 189 teragrams per year to 344 teragrams per year.[8]

Of these six contributors to other CO_2 emissions (iron and steel production, flaring of gas wells, cement production, natural gas systems, incineration of waste, and LULUCF), five are directly related to the design and operation of the built environment.

The emissions associated with the operation of natural gas systems are dependent on the combustion of natural gas for energy generation and heat production. Reductions in the need for these natural gas systems can be achieved through reductions in the energy use of buildings.

Two of these contributors are linked to the built environment through the embodied emissions of materials. The concept behind embodied emissions is that all materials require energy to produce—in the extraction of raw materials, processing, manufacturing, and transport between stages and to the point of use. It is difficult to derive the exact embodied emissions for building materials because of the variability in manufacturing processes and delivery distances for materials, but it is possible to make reasonable estimates. Refer to the case study in Figure 2.4. Two of the primary materials used to make the built environment are cement and steel, both of which are significant enough to appear as line items in the U.S. GHG inventory. These items are generally attributed to industry. Emissions reductions can be achieved through recycling of whole buildings, individual building elements, or materials, or by minimizing waste.

The built environment can also be a major source of waste, which in turn may be incinerated. Waste can occur during construction: approximately 10 percent of the material delivered to site is waste, through the need to cut stock sizes to fit or due to rework.[9] This can sometimes be reduced through prefabrication or prior coordination. Waste also occurs during building operation. In urban areas, the method of waste handling is usually determined by the local municipality and may include landfill, incineration, composting, or recycling, all of which have different CO_2 emissions associated with both the process itself and the required transportation. This is discussed further in Chapter 8.

Finally, the built environment can also form a part of the LULUCF carbon emissions source/sink. A change from undeveloped to developed land will generally result in a reduction of the amount of biomass stored, while the addition of trees to urban areas will result in an increase in the amount of biomass stored. Similarly the incorporation of "green infrastructure," such as bioswales, will result in a reduction compared to the use of "gray infrastructure." If urban areas are planned with climate change in mind, significant amounts of planting and biomass can be incorporated.

TABLE 2.2
Global warming potential for select GHGs
Data: Climate Change, 1995, p. 22[10]

Gas	Lifetime (years)	100-year global warming potential	Chemical formula
Carbon dioxide	variable	1	CO_2
Methane	12±3	21	CH_4
Nitrous oxide	120	310	N_2O
HFCs	1.5–264	140–11,700	multiple
Sulfur hexafluoride	3,200	23,900	SF_6
F-gases	2,600+	6,500–9,200	CF_4, etc.

CH_4 EMISSIONS

The next-largest set of emissions shown in Figure 2.1 is from CH_4. By far the largest contributor to these emissions is natural gas systems, followed by agriculture (cattle and sheep farming), landfills, and coal mining.

Of these, three can be affected by the design of the built environment. As noted above, if the natural gas systems required for energy generation can be made smaller, then the related emissions will also fall. The same applies to coal mining—if less is required, emissions will decrease. Landfills, like incineration, are one of the waste choices open to municipalities and are discussed further in Chapter 8.

N_2O EMISSIONS

By far the largest contributor to N_2O emissions in the United States is agricultural land management, with vehicle emissions running a very distant second. There is no significant direct impact to be made on N_2O emissions through the design of the built environment.

F-GAS EMISSIONS

The final group of emissions is F-gases. These gases are all human-made and used for industrial processes. The amounts of gas involved are all very small, but their effect is significant because they persist in the environment for very long periods and because their relative global warming impact is high. Relative impacts can be seen in Table 2.2.

Most of the impact is due to HFC and PFC gases that substitute for gases that were banned under the Montreal Protocol, such as CFCs and HCFCs, due to their ability to deplete the ozone layer. These gases are used as refrigerants, for foam blowing, as propellants, and in semiconductor manufacture, and they include all of the HFCs, along with R-22, a very common HCFC refrigerant that is still being phased out under the Montreal Protocol.

The most potent GHG is sulfur hexafluoride (SF_6), with a 100-year global warming potential of 23,900. It is used as an electrical insulator in large transmission switches and to inhibit oxidation in magnesium production. SF_6 emissions have fallen by nearly 50 percent in the last twenty years due to efforts by industry to reduce them, and this is a continuing trend.

Changes in the built environment that can decrease F-gas emissions include reducing refrigeration requirements and the selection of refrigerants with low global warming potential. The ways in which cooling requirements can be reduced in buildings are discussed in Chapters 7 and 9.

GREENHOUSE GAS EMISSIONS IN THE BUILT ENVIRONMENT IN SUMMARY

It is clear that the planning and design of the built environment have a significant effect on annual carbon emissions. In addition to the approximately 40 percent of carbon emissions directly associated with residential and commercial buildings by the GHG inventory, the design and operation of the built environment is a factor in the emissions associated with the following:

- vehicle travel, especially within urban areas
- waste, related particularly to whether it is recycled, incinerated, or sent to landfills
- refrigerants used for the generation of cooling in buildings
- operation of the natural gas supply system
- embodied energy of materials
- carbon contained within urban biomass.

Although not immediately apparent from the GHG inventory, the processing and delivery of water to buildings and the processing of wastewater is also an aspect of the built environment. An indication of the link between water, waste, energy, and carbon emissions is shown in Figure 2.6. The amount of energy needed to bring clean water to buildings varies by whether the water is available locally, available locally but needs desalination, or imported over large distances. When designing processes that can use water or energy in a variety of ways, it is important to remember that water has energy use and carbon emissions associated with it. The specific effects of buildings and the built environment on overall carbon emissions are discussed in Part 2.

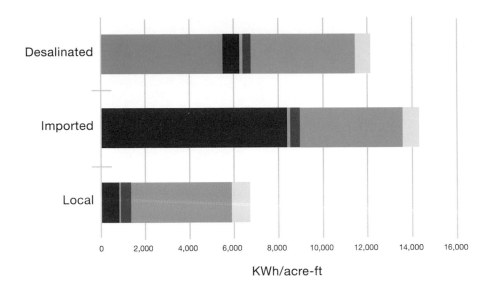

KWh/acre-ft

2.6 Indicative values for the amount of energy needed to bring clean water to buildings
Graphic: Arup

Legend

Wastewater treatment	Distribution	Conveyance
End use	Water treatment	Source

NOTES

1   Some 1,600 of the world's leading scientists, including the majority of the living Nobel laureates in the sciences, issued this appeal in November 1992. See http://fore.research.yale.edu/publications/statements/union.html.

2   U.S. Environmental Protection Agency (EPA), *Inventory of US Greenhouse Gas Emissions and Sinks: 1990–2009*, April 15, 2011, http://www.epa.gov/climatechange/emissions/usginventory.html (accessed June 5, 2012).

3   Nitrous oxide (N_2O) is one of the nitrogen oxides (NO_x) and the only one that has a GHG effect. Only 1 or 2 percent of the NO_x produced by combustion is N_2O, and most of it is nitric oxide (NO).

4   US Energy Information Administration, *Natural Gas 1998: Issues and Trends*, 1998, http://www.eia.gov/oil_gas/natural_gas/analysis_publications/natural_gas_1998_issues_and_trends/it98.html (accessed June 5, 2012)

5   See note 2 above.

6   United Nations Framework Convention on Climate Change, "Emissions from Fuel Used for International Aviation and Maritime Transport (International Bunker Fuels)," http://unfccc.int/methods_and_science/emissions_from_intl_transport/items/1057.php (accessed June 5, 2012).

7   As Amory Lovins has pointed out: "Of the energy in the fuel [an automobile] consumes, at least 80 percent is lost, mainly in the engine's heat and exhaust, so that at most only 20 percent is actually used to turn the wheels. Of the resulting force, 95 percent moves the car, while only 5 percent moves the driver, in proportion to their respective weights. Five percent of 20 percent is one percent—not a gratifying result from American cars that burn their own weight in gasoline every year." http://move.rmi.org/markets-in-motion/case-studies/automotive/hypercar.html (accessed June 5, 2012).

8   UNFCCC Consultative Group of Experts on National Communications from Parties not Included in Annex I to the Convention, *Handbook on Land-Use Change and Forestry Sector*, n.d., UNFCCC, http://unfccc.int/resource/cd_roms/na1/ghg_inventories/english/5_lucf/b_handbook/GHG_Inventory_in_Land_Use_Change_and_Forestry_Sector.doc (accessed June 26, 2012).

9   "In the UK an average of 13% of all materials delivered to site go into the skip without ever being used" (Scottish Environmental Protection Agency/CIRIA, *The Small Environmental Guide for Construction Workers*, CIRIA/SEPA, 2006, http://www.sepa.org.uk/customer_information/idoc.ashx?docid=f4881fdf-f986-480e-b023-b4ae1480e2ac&version=-1 [accessed June 5, 2012]).

10 Climate Change, *The Science of Climate Change: Summary for Policymakers and Technical Summary of the Working Group I Report*, 1995, http://unfccc.int/ghg_data/items/3825.php (accessed June 5, 2012).

i   IPCC, *Fourth Assessment Report of the Intergovernmental Panel on Climate Change*, New York: Cambridge University Press, 2007.

ii Oil: Dimensions Guide, "Barrel Dimensions," http://www.dimensionsguide.com/barrel-dimensions/ (accessed June 5, 2012); wiseGEEK, "Why Do We Measure Oil in Barrels?" http://www.wisegeek.com/why-do-we-measure-oil-in-barrels.htm (accessed June 5, 2012); U.S. EPA, "Calculations and References," http://www.epa.gov/cleanenergy/energy-resources/refs.html (accessed June 5, 2012). Aluminum, steel, hardwood: internal Arup research. Drywall: David Britz, Yoshi Hamaoka, and Jessica Mazonson, "Recology: Value in Recycling Materials," Cambridge, MA: MIT Sloan Management, http://actionlearning.mit.edu/files/slab_files/Projects/2010/report,%20recology.pdf (accessed June 5, 2012).

Highly detailed planet Earth at night, lit by the rising sun, illuminated by light of cities, surrounded by a luminous network—an informatics representation of the major air routes based on real data

Image: Anton Balazh/Shutterstock.com

3

Policies to Mitigate Climate Change

Fiona Cousins with Stephanie Glazer

Anything that is in the world when you're born is normal and ordinary and is just a natural part of the way the world works. Anything that's invented between when you're fifteen and thirty-five is new and exciting and revolutionary and you can probably get a career in it. Anything invented after you're thirty-five is against the natural order of things.

Douglas Adams, *The Salmon of Doubt*, p. 95[1]

INTERNATIONAL FRAMEWORK FOR POLICY

The generation of greenhouse gases (GHGs) is diffuse and worldwide, and pervades almost every industry and human activity. Concentrations of atmospheric GHGs can be increased through fuel combustion, deforestation, waste disposal, farming methods, and industrial production of fluorinated gases. The great variety of ways to affect GHG emissions is described in Chapter 2, which also describes the relative importance of different types of emissions.

A global problem like climate change needs a global solution, and the United Nations Framework Convention on Climate Change (UNFCCC) recognizes the need for action to limit average global temperature rise, the need for international cooperation, and that the greatest effects of the damage likely to be caused by climate change will be felt by those least able to manage it and least responsible for it. It also notes that countries have "common but differentiated responsibilities and respective capabilities" and that agreement must be made on the basis of equity.[2] These principles provide the platform for international negotiations to set emissions targets and share experience, but the Framework does not itself set specific targets. The parties to the convention meet annually at the Conference of Parties (COP) meetings to review the national communications of the participants and to discuss issues related to climate change. A number of richer, developed nations, known as Annex I countries, are responsible for most of the funding of the UNFCCC and are also responsible for most of the mitigation actions.

Kyoto Protocol

At the third COP meeting in Kyoto in 1997, a protocol was developed that included binding targets for emissions reductions. Before this could come into effect, it required ratification by at least 55 governments, with a minimum total commitment to affect at least 55 percent of the 1990 level Annex I countries' GHG emissions. The United States, emitting 36 percent of Annex I countries' global emissions, refused to ratify it in 2001, but the treaty was finally concluded in 2005 and presently has 192 signatories, including the European Union (EU) and the EU members.

The Kyoto Protocol is the only document that sets binding targets for national emissions reductions. It was intended to be the first of a series of agreements covering successive time periods, and its final compliance period ends in 2012. Following the principle of "common but differentiated responsibilities," every country has a different reduction target, and together these will lead to a 5 percent reduction in carbon emissions compared to the 1990 total for the Annex I countries. Only Annex I countries have emissions reductions targets—importantly, this excludes China, India, and Brazil.

In general the required emissions reductions are expected to take place within national boundaries, but there are three market-based mechanisms that allow countries to trade emissions reductions:

- Emissions trading, which allows countries to trade any "excess" reductions with countries that have not met their targets.
- Joint implementation, which allows developing countries to invest in emissions reductions in other developing countries but count the emissions reductions within their own targets.
- The clean development mechanism, which allows developed countries to implement an emissions reduction project in a developing nation.

As noted above, the UNFCCC recognizes that developing countries are likely to be most damaged by and least capable of managing the consequences of climate change. These three mechanisms explicitly support flows of both money and low-emissions technology from developed to developing nations.

The reporting of emissions is carefully prescribed by the protocol, and there are detailed rules in place to ensure *additionality*—that emissions reduction projects provide real reductions that would not otherwise have taken place.

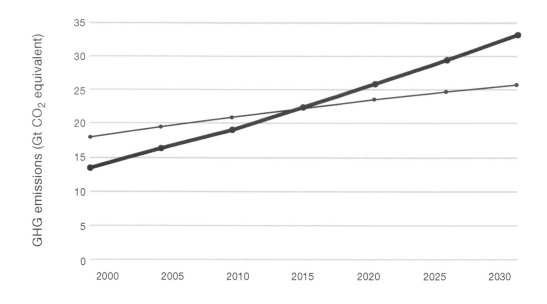

3.1 Greenhouse gas emissions by region over time
Graphic: Arup, Data: EPA[i]

Legend

Developed countries
Developing countries

As a binding agreement, the Kyoto Protocol has driven much national policy in those countries that have ratified it, and that have set targets, since its negotiation.

After Kyoto

It has proved very difficult for the COP to negotiate an extension to the Kyoto Protocol. There are many reasons for this, including the United States' failure to ratify it, Canada's withdrawal from it, and Japan's lack of appropriate policy to achieve its goals—all of which significantly weaken the treaty because these three countries accounted for nearly 40 percent of Annex I carbon emissions in 1990, the baseline year for the treaty.

In addition, there has been enormous growth in emissions in developing nations that don't have Kyoto Protocol commitments, in particular in China, India, and Brazil. The UNFCCC explicitly recognizes that developing nations have the right to develop and that this is more important than reducing carbon emissions. This recognition of development as a priority puts even more emphasis on rich countries' emissions reductions, as it is expected that developing nations' emissions will rise. As shown in Figure 3.1, it is expected that developing country emissions will exceed developed country emissions by 2015.

Emissions reductions targets that do not include China, India, and Brazil will likely not achieve global reduction goals as these countries compete with developed nations for manufacturing business. There is great concern in developed nations that any carbon emissions policy enacted only in developed nations will result in relocation of manufacturing facilities from developed to developing nations. This will make the policy ineffective—the emissions will not be reduced but will move to a different country, a process called *emissions leakage*—and will also put increased and potentially unfair pressure on the industries and the economies of developed nations. The issue of unfair competition was the stated reason behind George W. Bush's refusal to ratify the treaty, although this is usually seen as a failure of leadership by the United States.[3]

There was great hope that there would be progress toward an agreement at Copenhagen in 2009 (COP15), with less optimism for Cancún in 2010 (COP16) and Durban in 2011 (COP17).

Under the terms of the Copenhagen Accord, all countries, not just those in Annex I, were invited to make emissions reductions pledges to occur by 2020, with a baseline of 2005 emissions. Both China and India have made commitments related to *carbon intensity*—the amount of carbon emissions per unit of gross domestic product—rather than absolute reductions. China's pledge is 40 percent to 45 percent and India's is 20 percent to 25 percent. The United States' pledge is an absolute reduction of 17 percent. According to modeling by the Organisation for Economic Co-operation and Development (OECD), these pledges make up about 40 percent of the changes that the

Panel A: Annex 1

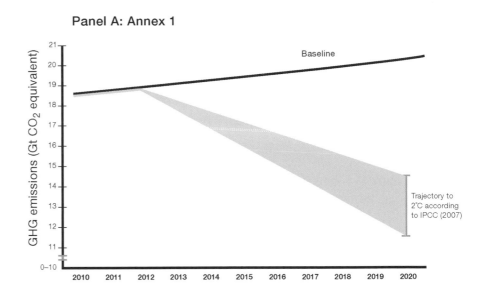

Panel B: World

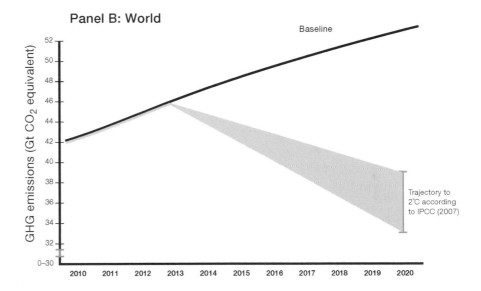

3.2 The gap between Copenhagen Accord pledged emissions reduction targets and the long-term goal of limiting global temperature increases to 2°C
Graphic: Arup, Data: OECD, 2010[ii]

Intergovernmental Panel on Climate Change has calculated to be necessary to hold temperature change at 2°C, which is the upper limit agreed under the UNFCCC (Figure 3.2).[4]

At Durban in 2011 (COP17), an agreement (the Durban Platform for Enhanced Action) was reached that goes some way to resolving these issues—a result far better than expected by most observers. Key headlines from the agreement are as follows:

- The Kyoto Protocol will extend into a second compliance period, ending in 2017 or 2020 (to be confirmed at COP18 in Qatar in 2012). This affects those countries with targets under Kyoto.

- A legally binding agreement on emissions that will affect all countries will be made by 2015, with an effective date of 2020.[5]

The weakness of the Durban outcome is that, while a commitment was made to a legally binding agreement, this is not the same as a commitment to set legally binding *targets*, so there is plenty of room for negotiation over the next three years. It is not at all certain that the consequent agreement will achieve emissions reductions that will hold global temperature rise to less than 2°C, let alone the 1.5°C that would be preferred by small island nations.

MAJOR POLICY MECHANISMS

In making policy to reduce carbon emissions, it is important to consider both the side effects of those policies and their effectiveness. Good policy will achieve the desired result, at the lowest cost, at the least inconvenience, and without negatively affecting the international competitiveness of the industry or country.

There are two major carbon emissions policy mechanisms open to governments: cap-and-trade and carbon taxation. Some countries use a mixture of the two, with different measures imposed on different industries. Both mechanisms may need additional policies to ensure that the overall costs are kept low enough not to impact competitiveness.

Cap-and-Trade

Cap-and-trade schemes calculate the maximum amount of emissions (the cap) allowed for a nation, industry, or sector. The total is divided into conveniently sized blocks of emissions allowances that are either auctioned in an open market or allocated to emitters at no cost, or some combination of the two. If a regulated emitter buys or is allocated more allowances than it needs, these allowances can be sold to emitters that do not have enough (the trade). Some cap-and-trade schemes allow carbon offsets from nonregulated sectors to participate in the emissions trading market as a cost-effective means for reducing the total GHG emissions attributed to the emitter.

This model was developed by the U.S. Environmental Protection Agency (EPA) and was extremely successful in reducing "acid rain" emissions at minimum cost. Clearly, cap-and-trade schemes are easier to implement when there are a limited number of emitters, and so they are usually applied to a specific industry. The success of a cap-and-trade scheme depends on setting the cap to encourage mitigation at reasonable costs and the means by which the emissions allowances are distributed. Cap-and-trade is the model used by the European Union Emissions Trading Scheme (EU ETS) and the Regional Greenhouse Gas Initiative (RGGI).

Carbon Taxation

In a carbon-tax scheme, a price is associated with the carbon content of fossil fuels and this price is paid when the fuels are purchased. Economists describe this as a tax on *externalities*. Carbon taxation

Counting Carbon

The key principles that provide the foundation for carbon accounting—relevance, completeness, consistency, accuracy, and transparency—require that double counting of emissions be avoided in order to have a valid emissions accounting. To this end, the World Resources Institute and World Business Council on Sustainable Development protocols for carbon accounting divide the emissions into three groups:

- **Scope 1**: All direct GHG emissions (with the exception of direct carbon dioxide [CO_2] emissions from biogenic sources)
- **Scope 2**: Indirect GHG emissions associated with the consumption of purchased or acquired electricity, steam, heating, or cooling
- **Scope 3**: All other indirect emissions not covered in Scope 2, such as emissions resulting from the extraction and production of purchased materials and fuels, transport-related activities in vehicles not owned or controlled by the reporting entity, outsourced activities, waste disposal, and others.

Note that this division is based on accounting by the corporation or entity, and that one entity's direct emissions (e.g., a power plant) may be another entity's indirect emissions (e.g., end user of electricity). The division of emissions into these scope categories mitigates the issue of double counting by providing a mechanism for an entity to claim unique control of direct emissions, while still taking responsibility for indirect emissions, even though they are directly emitted by another entity.[6]

This protocol is sometimes extended to carbon accounting for individuals. This can lead to some sense that there is a gray area—personal car commuting emissions are clearly Scope 1 for the individual and Scope 3 for the employer, but it may be difficult for individuals to mitigate these emissions if the workplace is inaccessible by other modes of transport.

We often try to extend this method of carbon accounting to buildings and campuses, but this is very difficult to do rigorously where the responsibility for building emissions is split between tenants or between the owner and tenants. The basic principles can be applied, but for buildings and campuses it is often more useful to set a boundary across which energy/carbon flows and to measure those flows.

makes carbon-intensive activities more expensive, which encourages fuel switching and reduced consumption. A carbon tax has a more pervasive effect on the whole economy than a cap-and-trade system and can result in lower overall implementation costs. It is often more difficult to implement politically, although the total tax burden does not have to increase.

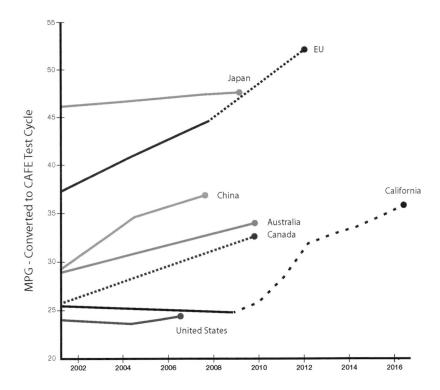

3.3 Comparison of fuel economy standards, normalized by CAFE converted miles per gallon. Dotted lines show projected rates
Graphic: Arup, Data: An and Sauer, 2004 [iii]

Comparison

Cap-and-trade almost guarantees the desired environmental outcome, but the costs imposed on emitters are not completely predictable as they depend on the total cap set, the cost of mitigation measures, and the speed with which those measures can be implemented—all of which can lead to price volatility for allowances and offsets. Unlike cap-and-trade, carbon taxation makes the costs for the consumer very clear because the price is fixed, but there is no guarantee that the wished-for emissions reductions will occur. Of course, modeling can help predict both the cost impact of a cap-and-trade system and the likely emissions reductions under a taxation scheme.

To compare the two systems, consider the following example: if electricity generation is subject to a cap-and-trade scheme, then it is likely that electricity prices will rise. If gas and oil are not also subject to a cap-and-trade mechanism, then electricity will become relatively more expensive than gas, driving consumers toward increased gas use. If the increase in electricity price—the revenue from cap-and-trade—is paying for decarbonization of the grid, this trend toward natural gas would be a poor outcome. This situation would not arise under a carbon tax regime because the price of electricity would fall as the grid decarbonizes.

Cost Containment

Both schemes can lead to revenue generation for the government. This revenue is usually used to fund other activities that reduce emissions or contain costs, and these activities need to be included in the policy. Examples include financing research or technology transfer to help reduce the cost of carbon emission reductions;

exemptions from employer health or social security costs for emitters that achieve emissions targets, thus making the emissions reductions less expensive, or even cost neutral, for the emitters; tax credits (for the tax scheme only); and allowing the use of offsets purchased from other industries or sectors for which carbon mitigation is easier.

Costs can also be contained and emissions leakage can be reduced using import taxes for high embodied-carbon goods from countries with no emissions targets. These tariffs need to comply with World Trade Organization guidelines.

Other Considerations

The rules around reporting carbon emissions are critical to long-term emissions reductions, especially when emissions allowances and offsets are to be traded. The standards must assure complete, accurate, comparable, reliable, and transparent reporting. This in turn will allow agreement on the fungibility of carbon credits and carbon emissions associated with offsets. Monitoring, verification, and reporting are also essential. Voluntary markets are built on robust, agreed accounting rules, while compliance markets also require significant work to develop and approve the policies.

There are other policies that can be used to promote specific behaviors to reduce carbon emissions without the broader framework of cap-and-trade or carbon taxation. These can be easier to negotiate by using arguments other than climate change, such as energy independence, air quality improvements, or cost reduction. Examples include building energy reduction legislation, as described later in this chapter, fuel efficiency standards for vehicles (Figure 3.3), or even "cash for clunkers" programs that aim to remove inefficient

Voluntary Action

In the absence of mandatory targets, many groups have tried to push for voluntary action. Examples include the films *Home*, by Yann Arthus-Bertrand, and Al Gore's *An Inconvenient Truth*, as well as work by 350.org.[7]

It can be difficult to make a coherent case for action for many reasons, but two of the main barriers are the separations in time and space between the causes and effects of climate change and the fact that the actions of individuals can seem so insignificant compared to the size of the issue.

Some of the most focused work on generating behavior change has been done by Columbia University's Center for Research on Environmental Decisions.[8] The framework for action that they have developed is outlined in Figure 3.4.

If you are designing a program to reduce carbon emissions, note that regular small changes in carbon emissions are generally more effective than occasional large ones, because these are habit-forming and cease to be an inconvenience. There is poor understanding about the relative importance of different actions, for example, changing lightbulbs compared to buying an efficient washing machine or changing cars.

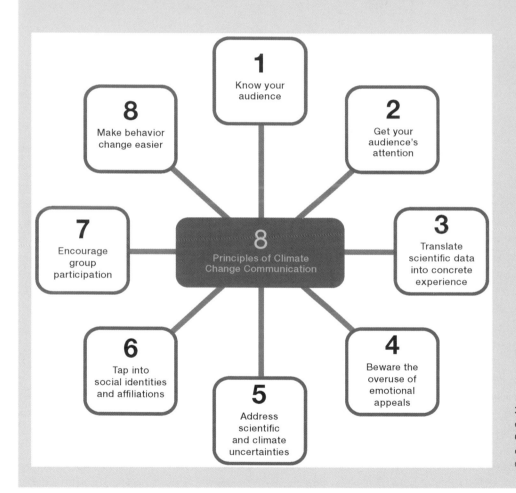

3.4 Eight principles of climate change communication
Graphic: Arup, Data: Center for Research on Environmental Decisions, 2009
Graphic: Arup, Data: An and Sauer, 2004[iv]

vehicles from the road.[9] The U.S. Energy Policy Act of 1992 was a policy of this type: it was intended to reduce carbon emissions in accordance with UNFCCC goals at no cost through energy efficiency and energy saving measures.[10]

POLICY EXAMPLES

United States

There is considerable pressure for carbon emission reductions and legislation driven by a number of factors, including a desire in industry for certainty around the cost of carbon into the future[11] and environmental worries, especially in regards to diminishing crop yields

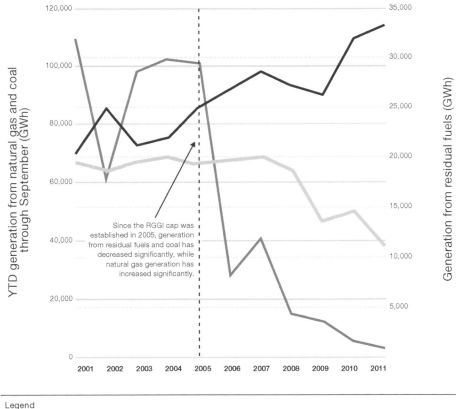

Since the RGGI cap was established in 2005, generation from residual fuels and coal has decreased significantly, while natural gas generation has increased significantly.

3.5 Changes in fuel mix for electricity generation showing the effect of the RGGI
Graphic: Arup, Data: ENE, 2011[v]

Legend — Natural gas — Coal — Residual fuel

and sea-level rise in a warming world. The United States has also pledged a 17 percent emissions reduction under the Copenhagen Accord.

However, cynicism about the reality of climate change together with a fear of increased taxation and concern about international competitiveness have made it hard to legislate at the federal level.[12] The Senate has considered a number of proposals to establish a cap-and-trade system or other related market mechanisms without success, most recently in 2010. The EPA has been working to implement new requirements under the Clean Air Act, following the Endangerment Finding,[13] which determined that GHG emissions are harmful to human health and the environment, and therefore requires the EPA to regulate emissions. While the legislature is not currently considering direct regulation of GHG emissions, the EPA continues to review and update its requirements under the Clean Air Act. Ongoing challenges to EPA's authority are being pursued by those opposed to regulation, although it was still in effect in early 2012.

These pressures have led to the adoption of various types of policies across the United States, usually at a regional, state, or city

level. Examples include the RGGI and California's Assembly Bill 32 (AB32), described below, as well as many local plans that impact emissions within cities, such as PlaNYC 2030 and the code changes promoted by the New York City Green Codes Task Force.

Regional Greenhouse Gas Initiative

The first functioning cap-and-trade system for carbon emissions mitigation in the United States was the RGGI, a collaboration of nine northeastern states intended to reduce emissions from the power sector by 10 percent by 2018.[14] Auctions are held regularly, and proceeds are returned to the states that invest in either low-carbon technologies or energy-efficiency programs. Many lessons have been learned from the RGGI since its inception on January 1, 2009. As with the EU ETS, the cap was initially set too high, but recent independent reports show that RGGI has generally been successful, as measured by carbon emissions reductions, job creation, customer price reductions through energy efficiency initiatives, and reduction in fossil fuel demand.[15] The change in fuel demand is shown in Figure 3.5.

Assembly Bill 32

At the state level, California has been a leader in setting aggressive regulatory targets. Its Global Warming Solutions Act of 2006 (AB32) requires that GHG emissions be reduced to 1990 levels by 2020. The state has committed to further reductions of 80 percent below 1990 levels by 2050. AB32 uses a variety of mechanisms to achieve emission reductions including a statewide cap-and-trade scheme, which covers the power sector and large industrial facilities (starting in 2012), and residential/commercial/industrial fuel combustion and transportation fuels (starting in 2015). California plans to participate in the Western Climate Initiative (WCI), an agreement similar to RGGI but applied throughout the economy, to help control emission reduction costs. WCI is due to go into effect in 2013, with the goal of reducing emissions by 15 percent from 2005 levels by 2020. The core WCI includes California, Manitoba, British Columbia, Ontario, and Quebec. In addition, AB32 identified a number of early action measures where legislation can result in specific emissions decreases that can be used as credits in the cap-and-trade scheme. These include landfills, ship electrification in ports, and fuel efficiency. Credit will also be available for voluntary early action measures.[16]

European Union Emissions Trading Scheme

The EU ETS is a cap-and-trade scheme designed to achieve the EU's Kyoto Protocol carbon emissions reduction obligations. It covers the 27 countries of the EU, plus Iceland, Liechtenstein, and Norway. The regulated industries are power generation, combustion plants, iron and steel production, and factories making cement, glass, lime, bricks, ceramics, pulp, paper, and board—which account for almost half of the EU's CO_2 emissions. Airlines,[17] petrochemicals, ammonia, and aluminum industries will join the scheme in future. Nitrous oxide emissions are also included.[18] At the end of each year, spare allowances can be kept for future years within the compliance period (so-called "banking") or traded to other companies. If a company does not have enough allowances and has not purchased allowances to cover its emissions at the end of a year, it may borrow against future years, up to the end of the compliance period, at which time it is subject to fine. Banking and borrowing was introduced to the EU ETS during Phase II of the scheme. The total number

of allowances will be 21 percent lower in 2020 than it was in 2005 when the scheme was launched.

Cost control was achieved by allocating emissions allowances at no cost in the first and second compliance periods (2005–7 and 2008–12). Allowances were assigned to companies according to a national allocation plan. For 2013–21, the next compliance period, at least half the allowances will be auctioned and allocation will be across the EU rather than to a national plan. Auctioning will become the rule for all sectors where emissions leakage is not considered to be an issue; electricity production allowances will be fully auctioned. Where free allowances are available, they will be provided by product (not sector) on the basis of the 10 percent most carbon emission efficient installations in the EU. The proportion of allowances auctioned each year will increase.

The third compliance period will also see the introduction of a 50 percent limit to access to project credits (joint implementation and clean development mechanism) from outside the EU. Revenues raised in the auction will allow the allocation of 12 percent of credits to poorer EU countries, and there is a nonbinding agreement that 50 percent of the revenue will be used for other climate mitigation issues.

The EU ETS has not been without problems. There was a collapse in the price of carbon in 2006 when it was discovered that too many allocations had been made in the first compliance period. In spite of this, there was a reduction in carbon emissions of about 2 percent by 2007. This issue was thought to be resolved in the second compliance period as the price of carbon was stable at approximately €13 per ton from 2008 to 2010. The global financial crisis of 2008 and subsequent recession meant that many companies reduced their emissions through reduced need for goods and have banked the allowances for future years; in some industries, no technological change will be needed until 2016.[19] These factors have led to a fall in carbon prices to approximately €8.5 per ton. This type of price response is as expected, and the scheme should bring about the desired emissions reductions.

China

China is not subject to a mandatory emissions reduction under the Kyoto Protocol but is the largest emitter in the world. It has pledged to reduce emissions intensity by 40 to 45 percent from 2005 levels by 2020 under the Copenhagen Accord. China's position on carbon emissions has shifted dramatically in the last few years, and there are now plans in place to reduce carbon emissions. The change in view is likely due to increased wealth and competitiveness with the West, coupled with a realization that global problems will have effects in China and that China has some power to reduce the effects.

China has a climate change legislation project underway that will investigate various policy initiatives for emissions reductions. Draft legislation is expected in 2012 that will support the objective of the twelfth five-year plan, covering 2011 to 2015, which requires the country to reduce the CO_2 emitted per unit of gross domestic product by 17 percent.[20] This is almost certainly a prelude to absolute emissions reductions legislation, and the optimistic view would be that this will assist the UNFCCC parties to come to a binding agreement on emissions in the future, as China will then have binding targets.

In January 2012, China's National Development and Reform Commission asked the cities of Beijing, Tianjin, Shanghai, Chongqing, and Shenzhen, and the provinces of Hubei and Guangdong, to set overall emissions targets with a view to establishing an internal carbon-trading mechanism.[21]

Australia

Australia is a small emitter in global terms but has the highest per capita carbon emissions in the world due to its reliance on coal for electricity generation. Australia has recently passed a package of climate change legislation that will put a price on carbon, starting in July 2012. The policy model is for a carbon tax, set at A$23 per ton and fixed for three years, with a transition to a carbon-trading scheme in 2015.[22]

As the model is a tax it will raise revenue, and this money is being recycled back into the economy in a number of forms. There are tax breaks for individuals as well as trade-exposed corporations, along with specific initiatives aimed at reducing carbon emissions, including subsidies for solar water heating. It is too early to measure the success of the legislation, but it is the result of 10 years of public debate, so has been thoroughly tested and discussed.

SPECIFIC POLICIES AFFECTING THE BUILT ENVIRONMENT

United Kingdom

The European Union Energy Performance of Buildings Directive (EPBD) was passed in 2002 with the intention of improving building energy efficiency throughout Europe. One of its primary actions has been to mandate energy labeling for buildings. Two energy labels are required: an Energy Performance Certificate, which rates the performance of the building as designed, and a Display Energy Certificate (Figure 3.6 overleaf), which rates the performance of the building as operated. Both labels are comparative and express their results in terms of carbon. They are also publicly available so that performance can be managed over time and taken into account on sale and rental.

In June 2010, an updated EPBD set targets for new public buildings to be nearly zero energy by 2018 and for all new buildings to reach that target by 2020. The UK has been working toward making all new buildings zero carbon since 2006, with targets of 2016 for housing and 2019 for nonresidential, using the logic that new buildings should not add to national carbon emissions. The goal is to achieve this without offsets, using only efficiency standards and renewables. The Building Regulations are setting tighter targets for regulated energy use every three years, and unregulated energy is being targeted through a Carbon Reduction Commitment, which requires building occupants to pay a surcharge based on the carbon emissions related to their annual energy bills.[23] This is already starting to change attitudes about fit-out standards and operational standards. The UK is also using long-term loans as a means for homeowners to pay for energy-efficiency improvements over time. Using surcharges and long-term loans is intended to help control the rebound effect, or Jevons paradox, under which money freed up by energy savings is frequently used to buy additional appliances that consume more energy.

The UK also has a building rating system, BREEAM, which covers a wide range of green building issues including energy and water use. Specific levels of achievement under BREEAM are sometimes used as part of planning requirements for new developments, which tends to push higher levels of energy efficiency than would be achieved under basic regulations.

3.6 Example Display Energy
Certificate
Graphic: Arup

Display Energy Certificate
How efficiently is this building being used?

A Government Dept
12th & 13th Floor
Jubilee House
High Street
Anytown
A12CD

Certificate Reference Number:
1234-1234-1234-1234

This certificate indicates how much energy is being used to operate this building. The Operational Rating is based on meter readings of all the energy actually used in the building. It is compared to a benchmark that represents performance indicative of all buildings of this type. There is more advice on how to interpret this information on the government's website: www.communities.gov.uk/epbd.

Operational Rating

This tells you how efficiently energy has been used in the building. The numbers do not represent actual units of energy consumed; they represent comparative energy efficiency. 100 would be typical for this kind of building.

More energy efficient

A 0–25

B 26–50

C 51–75

D 76–100

100 would be typical

E 101–125 ◀108

F 126–150

G Over 150

Less energy efficient

Total CO$_2$ Emissions

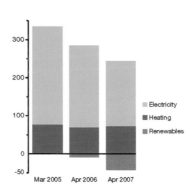

Electricity
Heating
Renewables

Mar 2005   Apr 2006   Apr 2007

Previous Operational Ratings

This tells you how efficiently energy has been used in this building over the last three accounting periods.

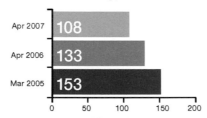

Apr 2007   108

Apr 2006   133

Mar 2005   153

0   50   100   150   200

Technical Information

This tells you technical information about how energy is used in this building. Consumption data based on actual readings.

Main heating fuel: gas
Building environment: air conditioned
Total useful floor area (m²): 2927
Asset rating: 92

	Heating	Electrical
Annual energy use (kWh/m²/year)	126	129
Typical energy use (kWh/m²/year)	120	95
Energy from renewables	0%	20%

Administrative Information

This is a Display Energy Certification as defined in SI2007:991 as amended.

Assessment software: OR v1
Property reference: 891123776612
Assesor name: John Smith
Assesor number: ABC12345
Accreditation scheme: ABC Accreditation Ltd
Employer/trading name: EnergyWatch Ltd
Employer/trading address: Alpha House, New Way, Birmingham, B21AA
Issue date: 12 May 2007
Nominated date: 01 Apr 2007
Valid until: 31 Mar 2008
Related party disclosure: EnergyWatch are contracted as energy managers
Recommendations for improving the energy efficiency of the building are contained in report reference number 1234-1234-1234-1234.

3.7 California and United States per capita electricity consumption (notice the flatlining of the California figures since the early 1970s)
Graphic: Arup, Data: California Energy Commission, 2004

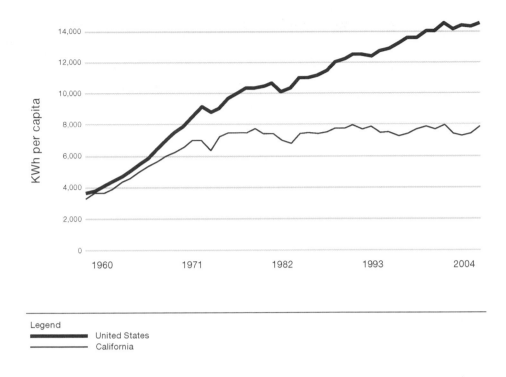

United States

In the United States, the major pieces of federal legislation that have affected the built environment have been the Energy Policy Acts (EPActs) of 1992 and 2005. These broad pieces of legislation include many sectors but have specific provisions for buildings. Perhaps the most important element is that EPAct 1992 established ASHRAE Standard 90.1 as the basis for all state energy codes and amended the Energy Policy and Conservation Act to give the Department of Energy (DOE) power to determine whether each new edition of this standard will provide improved energy efficiency in commercial buildings. If the DOE finds that it does, then the states are required to certify that their building standards are at least as stringent.[24] In 2011, the DOE determined that the 2010 standard would lead to an improvement in commercial energy efficiency, and states have until July 2013 to make their codes comply. EPAct 2005 also provided a package of tax incentive measures for commercial building energy efficiency improvements.

California has led the United States in emissions reductions with the adoption of AB32. It has also led the nation in energy efficiency standards, which are one of the key emissions reductions tools, through the adoption of Title 24 in 1978. Title 24 has always been more stringent than the ASHRAE standard, and this has led to a different trajectory for energy use in California compared to the rest of the country, as shown in Figure 3.7.

The LEED rating system was introduced by the U.S. Green Building Council in 1999 and has been regularly updated since. Like BREEAM,

it covers a broad range of green building issues, and one of its core components is energy modeling, using ASHRAE 90.1 as a baseline, to achieve points. LEED's adoption of updated versions of the code is generally ahead of the states' adoption, and LEED also requires a minimum performance of 15 percent better than the baseline. A number of cities have chosen to include a requirement for LEED certification in their building codes or for their own buildings, including San Francisco, Seattle, Houston, Boston, and New York City.

New York City has also been working to reduce carbon emissions as part of PlaNYC 2030, a plan developed to work out how best to accommodate 1 million more people within the city limits by 2030. At present, a number of energy efficiency, recycling, and other carbon emissions–related regulations are winding their way through the New York City Council.[25] These regulations include the first building energy efficiency and water use reporting in the United States and a requirement for regular building energy audits.[26]

China

Chinese policy for low-carbon buildings is new and not yet fully formed. The new requirements for emissions reductions (see the information on China in the "Policy Examples" section) will likely have some effect in the near future and, in a planned economy, things can move very quickly once an appropriate path is defined.

China has defined a voluntary 3-Star Green Building assessment system that covers similar ground to the LEED and BREEAM systems. This is beginning to become a prerequisite for some land deals and is likely to become more regularly used as building codes are more strictly enforced.

In the meantime, there are a number of "eco-cities" under development. They are not expected to be more than 15 percent better than the statutory minimum. The definition of *eco-city* is also under development.

POLICIES TO MITIGATE CLIMATE CHANGE IN SUMMARY

As shown, the policies selected by different governments vary widely. Those countries that have made overarching policy have used both cap-and-trade and taxation mechanisms. The cap-and-trade mechanisms are generally limited to specific industries so that they can more easily be managed. In addition to the overarching policies, there is usually related legislation specifically affecting different sectors, such as the EPAct in the United States and Building Regulations in the United Kingdom. This legislation is sometimes used to point the way to solutions that reduce the financial impact of cap-and-trade or taxation on consumers and is sometimes used to regulate other sectors where cap-and-trade would be unwieldy or would not produce the desired results.

Many governments have recognized that the built environment is a major source of carbon emissions and that these emissions can be tackled through existing technologies at moderate cost and with a positive return on investment. The specific measures that can be taken in the built environment and the contribution of the built environment to carbon emissions are discussed fully in Part 2, but it is important to note that many of these measures are mandated through legislation at national and local levels.

FURTHER READING

Houser, Trevor, Rob Bradley, Britt Childs, Jacob Werksman, and Robert Heilmayr, *Leveling the Carbon Playing Field: International Competition and US Climate Policy Design*, Washington, DC: Peterson Institute, 2008.

Leggett, Jane, Richard Lattanzio, Carl Ek, and Larry Parker, *An Overview of Greenhouse Gas (GHG) Control Policies in Various Countries*, Washington, DC: Congressional Research Service, 2009.

UNFCCC, *Report of the Conference of the Parties on Its Fifteenth Session, Held in Copenhagen from 7 to 19 December 2009*, UNFCCC, 2009, http://unfccc.int/meetings/copenhagen_dec_2009/meeting/6295/php/view/reports.php (accessed June 7, 2012).

UNFCCC, *Report of the Conference of the Parties on Its Sixteenth Session, Held in Cancún from 29 November to 10 December 2010*, UNFCCC, 2010, http://unfccc.int/meetings/cancun_nov_2010/meeting/6266/php/view/reports.php (accessed June 7, 2012).

NOTES

1   London: Macmillan, 2002.

2   United Nations Framework Convention on Climate Change, May 9, 1992, Art. 3, http://unfccc.int/essential_background/convention/background/items/1355.php (accessed June 7, 2012).

3   The Senate unanimously passed a bill promoted by Robert Byrd (D-WV) and Chuck Hagel (R-NE) in 1997 that instructed the administration not to sign any protocol that did not require developing countries to reduce their emissions, on the basis that it would be unfair and ineffective. This effectively prevented the president from signing on to Kyoto and set the outlines for more recent negotiation.

4   OECD (Organisation for Economic Co-operation and Development), *Costs and Effectiveness of the Copenhagen Pledges: Assessing Global Greenhouse Gas Emissions Targets and Actions for 2020*, Paris: OECD, 2010, http://www.oecd.org/dataoecd/6/5/45441364.pdf (accessed June 7, 2012).

5   "The Ad Hoc Working Group on the Durban Platform for Enhanced Action shall complete its work as early as possible but no later than 2015 in order to adopt this protocol, legal instrument or agreed outcome with legal force at the twenty-first session of the Conference of the Parties and for it to come into effect and be implemented from 2020" (UNFCCC COP17, "Establishment of an Ad Hoc Working Group on the Durban Platform for

Enhanced Action: Proposal by the President," draft decision, Durban, 2011, http://unfccc.int/files/meetings/durban_nov_2011/decisions/application/pdf/cop17_durbanplatform.pdf [accessed June 7, 2012]).

6   Local Government Operations Protocol, *For the Quantification and Reporting of Greenhouse Gas Emissions Inventories*, Version 1.1, May 2010. Developed in partnership and adopted by California Air Resources Board, California Climate Action Registry, ICLEI—Local Governments for Sustainability.

7   http://www.350.org/ (accessed June 7, 2012).

8   http://cred.columbia.edu/ (accessed June 7, 2012).

9   In the UK, vehicle tax follows a stepped scale related to carbon emissions per mile.

10 Parker, Larry, John Blodgett, and Brent Yacobucci, *US Global Climate Change Policy: Evolving Views on Cost, Competitiveness, and Comprehensiveness*, Washington, DC: Congressional Research Service, 2010.

11 Refer to the United States Climate Action Partnership for an industry-based call to action: http://www.us-cap.org/ (accessed June 7, 2012).

12 Senator James Inhofe: "I have offered compelling evidence that catastrophic global warming is a hoax. That conclusion is supported by the painstaking work of the nation's top climate scientists" ("Inhofe Delivers Major Speech on the Science of Climate Change: Catastrophic Global Warming Alarmism Not Based on Objective Science," press release, July 28, 2003, http://inhofe.senate.gov/pressapp/record.cfm?id=206907 [accessed June 7, 2012]).

13 EPA, "Endangerment and Cause or Contribute Findings for Greenhouse Gases under Section 202(a) of the Clean Air Act," http://epa.gov/climatechange/endangerment.html (accessed June 7, 2012).

14 Connecticut, Delaware, Maine, Maryland, Massachusetts, New Hampshire, New York, Rhode Island, and Vermont. New Jersey withdrew in 2011.

15 ENE, "RGGI Emissions Trends," Rockport, ME: ENE, January 2011, http://env-ne.org/public/resources/pdf/ENE_RGGI_Emissions_Report_120110_Final.pdf (accessed June 7, 2012); Paul J. Hibbard, Susan F. Tierney, Andrea M. Okie, and Pavel G. Darling, *The Economic Impacts of the Regional Greenhouse Gas Initiative on Ten Northeast and Mid-Atlantic States*, Boston, MA: Analysis Group, November 15, 2011, http://www.analysisgroup.com/uploadedFiles/Publishing/Articles/Economic_Impact_RGGI_Report.pdf (accessed June 7, 2012).

16 See http://www.arb.ca.gov/cc/ab32/ab32.htm.

17 The intention to apply the ETS to aviation was challenged by the U.S. airlines during 2011. The European Court of Justice upheld the legality of applying the ETS to international flights in December 2011.

18 European Commission, "Emissions Trading System," http://ec.europa.eu/clima/policies/ets/index_en.htm (accessed June 7, 2012).

19 Damian Carrington, "EU Emissions Trading Scheme on Course to Make Tiny Savings, Says Report," *The Guardian*, September 9, 2010, http://www.guardian.co.uk/environment/2010/sep/10/eu-emissions-trading-savings (accessed June 7, 2012).

20 "Key Targets of China's 12th Five-Year Plan," Xinhuanet, March 5, 2011, http://news.xinhuanet.com/english2010/china/2011-03/05/c_13762230.htm (accessed June 7, 2012).

21 "China Orders 7 Pilot Cities and Provinces to Set CO2 Caps," Reuters, January 13, 2012, http://www.reuters.com/article/2012/01/13/us-china-carbon-idUSTRE80C0GZ20120113 (accessed June 7, 2012).

22 See Australian Government Department of Climate Change and Energy Efficiency, http://www.climatechange.gov.au/.

23 Regulated energy includes lighting and equipment efficiencies. Unregulated energy includes discretionary items like monitors and computers.

24 U.S. DOE, "All You Ever Wanted to Know About Energy Code Determinations," http://www.energycodes.gov/status/all_about_determinations.stm (accessed June 7, 2012).

25 For further information, go to http://www.nyc.gov/html/planyc2030/html/home/home.shtml, http://www.urbangreencouncil.org/.

26 Refer to the Greener, Greater Buildings Plan: http://www.nyc.gov/html/planyc2030/html/about/ggbp.shtml.

i   http://www.epa.gov/climatechange/ghgemissions/global.html (accessed June 7, 2012).

ii See note i in Chapter 2.

iii An, F. and Sauer, A., *Comparison of Passenger Vehicle Fuel Economy and Greenhouse Gas Emission Standards Around the World*, Arlington, VA: Pew Center on Global Climate Change, 2004, http://www.c2es.org/docUploads/Fuel%20Economy%20and%20GHG%20Standards_010605_110719.pdf (accessed June 7, 2012).

iv Center for Research on Environmental Decisions, *The Psychology of Climate Change Communication: A Guide for Scientists, Journalists, Educators, Political Aides, and the Interested Public*, New York: Center for Research on Environmental Decisions, Columbia University, 2009.

v   ENE (Environment Northeast), *RGGI Emissions Trends*, Rockport, ME: ENE, January 2011, http://www.env-ne.org/public/resources/pdf/ENE_RGGI_Emissions_Report_120110_Final.pdf (accessed June 7, 2012).

Beddington Zero Energy Development is a live/work housing development in Hackbridge, London, England

Image: Arup Assetbank

4

Sustainability and Climate Change

Alisdair McGregor

A new scientific truth does not triumph by convincing its opponents and making them see the light, but rather because its opponents eventually die, and a new generation grows up that is familiar with it.

Max Planck[1]

Climate change has a direct impact on the environment and all living things on the planet. Severe economic and social impacts quickly follow the environmental impact. The relationship of these drivers is illustrated in Figure 4.1. The basic tenet of sustainability is to take actions now so that future generations can continue to enjoy standards of living similar to those today. Additionally, for future generations in much of the developing world standards of living need to improve significantly—as the previous chapters have shown, the economic and social impacts of climate change are likely to be most severe in developing countries.

SUSTAINABLE DESIGN

While most people would agree that climate change is a real problem and that something should be done to mitigate the worst scenarios, the sheer size of the problem leaves individuals, organizations, and in some cases countries at a loss as to what can be done. Sustainable approaches to policy, design, and management provide a framework for tackling global warming at the level of individual policies, projects, and corporate strategies.

The underlying approach to sustainable design is to consider environmental, economic, and social issues in a balanced way. Sustainable design provides an ideal framework for developing strategies to mitigate climate change, while balancing the environment, economy, and social well-being. The complexity of defining and quantifying the impacts and influences of any particular project and then comparing them to arrive at an ideal solution is daunting. Clearly, finding the optimal balance of environmental, economic, and social components is not an easy task. Several sustainable design tools

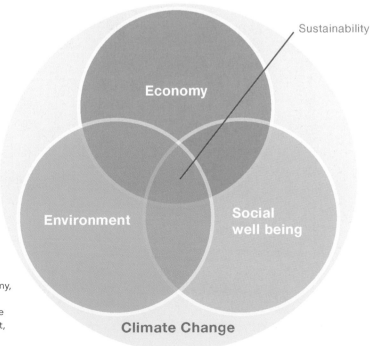

4.1 Climate change affects the economy, environment, and social well-being; sustainability addresses climate change while balancing economy, environment, and social well-being

Graphic: Alisdair McGregor

SPeAR

SPeAR is a holistic sustainability decision-making framework to support project development and communicate outcomes, in the form of a software-based tool for use on a wide variety of projects, across the globe.

It encompasses an integrated quantitative and qualitative appraisal based on a library of indicators that can be adapted to specific project conditions. Core indicators are presented graphically on the SPeAR diagram using a traffic light–type system (Figure 4.2) to indicate performance against key themes. The software also generates a tabulated summary of the input data, ensuring that the process is robust and auditable.

For more information, go to www.arup.com/Projects/SPeAR.aspx or see a SPeAR demonstration at www.oasys-software.com/spearapp/app/index.html.

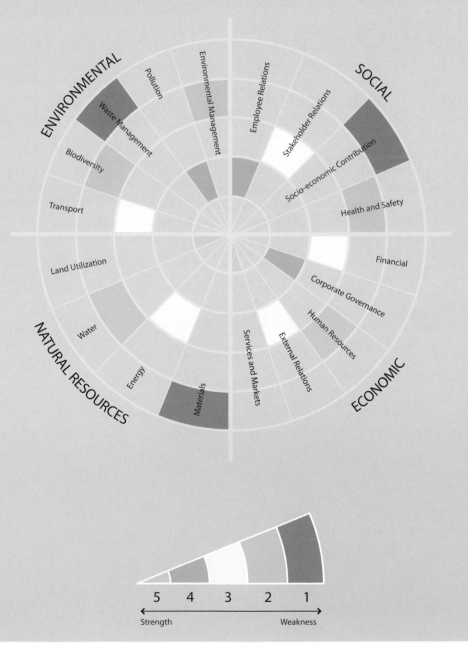

4.2 SPeAR diagram
Graphic: Arup

5 4 3 2 1
←——————————→
Strength   Weakness

Integrated Resource Management

The Integrated Resource Management (IRM) methodology evaluates interrelationships between and among masterplans, and how future developments will use resources. This methodology is used to develop models for use in large masterplanning projects around the world. There are a number of IRM implementations, each customized for development and client preferences or requirements.

These models allow planners to first understand the land use and resource impact of an existing or proposed development, and then evaluate sustainable development strategies for that development. The impacts of these strategies can be observed in terms of energy use, carbon emissions, water use, waste and wastewater generation, vehicle miles traveled, food consumption, and other sustainability indicators.

Arup collaborated with Waterfront Toronto, the Climate Positive Development Program, and the University of Toronto in the first test case of this model for Waterfront Toronto's West Don Lands project—a 32-hectare (80-acre) redevelopment site east of downtown Toronto that is part of one of the largest waterfront redevelopment initiatives ever undertaken in the world.

Arup's pilot model enables the agencies to quantify the sustainability performance of their redevelopment projects—the energy, water, waste, materials, and greenhouse gas emissions related to development decisions.

Ultimately, the tool is helping to accelerate a low-carbon future by providing city planners and large-scale developers with the tools they need to understand and address the greenhouse gas emissions generated by their developments and "get to zero."

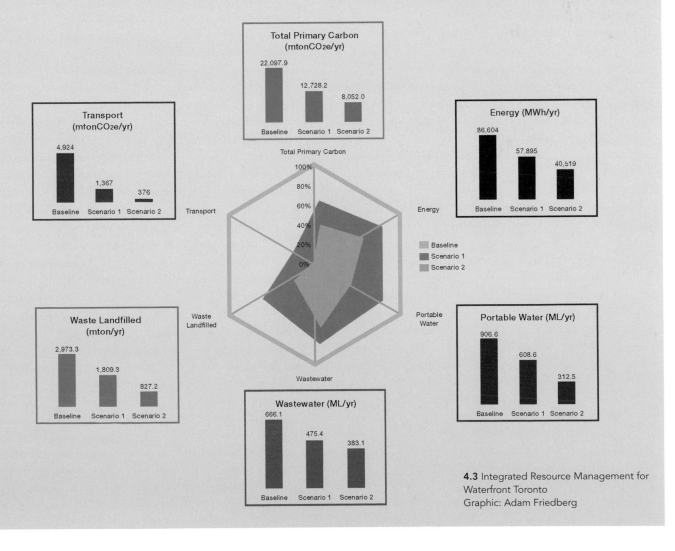

4.3 Integrated Resource Management for Waterfront Toronto
Graphic: Adam Friedberg

have been developed to help analyze and balance the competing influences. Some tools rely on developing key performance indicators for ranking different strategies. Arup's Sustainable Project Appraisal Routine (SPeAR) is an example of this type of tool (see box).[2] Alternate tools compare the carbon footprints of different planning, transit, and energy strategies. The Integrated Resource Management (IRM) tool developed by Arup is an example of this type of tool (see box). These tools are aimed at developing a holistic understanding of this complexity. The "Mitigation" and "Adaptation" sections of this book demonstrate ways to arrive at balanced solutions.

This integrated approach is essential to getting around the most common argument against mitigation strategies—that taking appropriate action now will harm the economy. This requires a different approach to economic analysis, such as that seen in the Stern Review (see also Chapter 12).[3] At the project level, there is a need to move away from simple payback analysis of energy costs saved toward a value-based system. Value has to be established for metrics such as improved occupant satisfaction, corporate image, and reduced absenteeism. In many cases, cost perception rather than real cost is the stumbling block. By taking a broader look at the long-term value of a project rather than merely the short-term payback, it is possible to demonstrate that a more sustainable approach is the best option.

Sustainable design demands a holistic and balanced approach; when applied correctly, it goes beyond climate mitigation strategies that ignore social and economic issues. Without this holistic approach, many of the strategies proposed for combating climate change founder as economic and social development advocates are pitted against environmentalists. The balanced, win-win-win approach can be illustrated in the planning of sustainable communities. Individual systems and strategies can be pulled out as uneconomic, but when sustainable design is viewed from a holistic standpoint, a rich, vibrant community can result, coinciding with thriving economic development. The Beddington Zed project in the United Kingdom is a good example.[4] This close-to-zero-energy housing development in the South of London was successful because people wanted to live an affordable, eco-friendly lifestyle. So the value of the project is increased. This is an essential part of the social leg of sustainable design. People should want to live and work in low-carbon buildings and communities because they are beautiful and great places to be. Austerity will not sell enough low-energy buildings to meet the carbon reduction needed over the next few decades. Buildings, places, and cities that are well loved will be maintained better and probably have a much longer lifetime. Long life is in itself a climate mitigation strategy, as the embodied carbon emissions are spread out over a much greater time period.

Just as damaging as the perception that sustainable designs are always expensive is the assumption that sustainable design principles will override aesthetics. The sustainable design framework in fact demands the opposite. It is essential that buildings and products that have a low carbon footprint also provide joy and functionality to those who purchase them. This places a lot of responsibility on the designers and constructors of the built environment. To achieve just low- or zero-carbon performance is not sufficient. Beauty and increased functionality must also be integrated into the solution. This concept can be illustrated by an example from the auto industry. The Honda Insight was one of the first hybrid vehicles introduced. Its fuel economy was very impressive, but it was barely a two-seater, with limited functional appeal to the typical family. As for its aesthetic—well, it certainly wouldn't pull the crowds at an auto show. The Toyota Prius had less impressive fuel consumption numbers but was a comfortable four-seater with good luggage space and, while maybe not the most stylish car, had a more acceptable aesthetic. The Prius went on to become a very successful product line, resulting in a far greater net carbon reduction than the more fuel-efficient Insight. Sustainability provides the framework to balance aesthetics and pure carbon performance.

The good news is that sustainable buildings tend to have a greater market appeal than business-as-usual ones. Survey work from the Center for the Built Environment shows that occupants give higher satisfaction ratings for LEED-rated and other recognized green buildings than for standard buildings.[5] Reviews of sustainable communities show that these developments sell faster and have a higher market value than standard developments. In Europe, compact transit-oriented communities are almost the norm, but in the United States, where land has been plentiful, suburban sprawl has been the American dream. Survey reports are starting to show that home buyers value community and walkable neighborhoods more than house size. Most Americans would choose a smaller home and smaller lot if it would keep their commute time to 20 minutes or less.[6] There is also an increasing understanding of the value of sustainable infrastructure for community development.[7] These issues are covered in greater detail in Chapter 8.

Achieving the balance between the many components of sustainability requires a fully integrated team. Chapter 10 covers integrated design in greater detail. Team members on a sustainable project must have a wider understanding of overall drivers and impacts of the project—specialists have to see how their respective disciplines fit into the overall project. Successful sustainable design requires abandoning most of the traditional linear approaches to design and construction, replacing them with a holistic approach. The short time frame in which the carbon emissions associated with the built environment must be reduced means that the integrated team approach has to be introduced rapidly. Some architects will have to give up their positions as sole leaders of design efforts. This upheaval of the traditional approach will result in low-carbon projects that are economic, environmentally sound, and beautiful.

EDUCATION FOR SUSTAINABILITY AND CLIMATE CHANGE

This chapter has argued that sustainable design is the ideal framework for responding to climate change and that multidisciplinary teams are essential to developing successful solutions. The question remains, are there sufficient numbers of people being educated to form these teams? Edward Mazria, the author of the 2030 Challenge in the United States, was greatly concerned by a survey of American architectural schools which showed that not one required classes in sustainable design. While many offered optional classes in sustainable design, the fact that it was possible to graduate with an architectural degree without covering any sustainable topics was worrying. In 2007, Mazria launched the 2010 Imperative: Global Emergency Teach-In with the goal of making sustainable design classes a basic requirement in all architecture schools by 2010.[8] Although this might seem an ambitious target, failure to meet it makes achieving carbon neutrality in all new buildings by 2030 even more difficult. While not completely meeting the 2010 target, sustainable design and policy

courses are gradually making their way into higher education, but with the time lag until these newly educated students reach positions of influence, other steps must be taken immediately.

ORGANIZATIONAL SUSTAINABILITY

On a larger scale, sustainability again provides the organizational framework for corporations and organizations to respond to climate change. Just as in the design of the built environment, sustainability brings a balanced approach to management that seeks the optimum balance between economic, environmental, and social considerations. This is a shift from the traditional way of measuring companies solely by their financial performance to considering a triple bottom line of accounting. We are now seeing companies using such reporting methods as the Balanced Scorecard or the Global Reporting Initiative.[9] Reporting methodologies like these reflect a move away from short-term, profit-focused corporate strategy. Changing to a more balanced and longer-term outlook is an essential first step in efforts to mitigate climate change.

It is not just the boutique, obviously green companies that are adopting a sustainable strategy. Many of the most successful Fortune 500 companies are setting bold targets for themselves. The common thread with these organizations is the recognition that sustainable, low-carbon business models are the best long-term business plan. The leadership displayed by these corporations is essential; however, government policy needs to reward leadership and raise minimum performance levels so that all companies are operating at a more sustainable level. In contrast, the European Union is setting the agenda for sustainable policy in cooperation with business. Corporate leadership and reporting is covered in greater detail in Chapter 13. Walmart is reviewed as an in-depth case study in Chapter 14.

DEVELOPING NATIONS

As stated at the start of this chapter, much of the largest impact of climate change will be felt in the developing nations. These nations need the greatest improvement in economic and social standing. Some of them stand at the tipping point of economic and environmental collapse. In his book *Collapse: How Societies*

Choose to Fail or Succeed, Jared Diamond examines many of these nations.[10] His review of the two adjacent nations of Haiti and the Dominican Republic illustrates how the sustainable environmental policies of the Dominican Republic have led to better economic and social conditions than those in neighboring Haiti. Because of the pressing need to improve the economies of developing nations, governments and developers tend to adopt short-term solutions rather than balanced long-term solutions. Generating sustainable frameworks and demonstrating that they will work for developing nations remains one of the greatest challenges in combating climate change.

SUSTAINABILITY AND CLIMATE CHANGE IN SUMMARY

Sustainable thinking applied to policy, planning, and design provides the framework for tackling the impacts of climate change in a balanced way. The sustainable approach demonstrates the value added when we act immediately in response to climate change. In other words, the sustainable option can clearly be considered something that people *want*, rather than merely something they ought to have. As we move deeper into the twenty-first century, we will see greater social and financial pressure to take more action to mitigate the impacts of global warming. This force is already driving policy and design. But can we change fast enough? The climate is not going to wait for us.

NOTES

1   *Scientific Autobiography and Other Papers*, trans. F. Gaynor, 1950, 33.

2   See http://www.arup.com/Projects/SPeAR.aspx.

3   *Stern Review: The Economics of Climate Change*, 2006, http://webarchive. nationalarchives.gov.uk/+/http:/www.hm-treasury.gov.uk/independent_ reviews/stern_review_economics_climate_change/stern_review_report. cfm (accessed June 7, 2012).

4   BioRegional, "BedZED, UK," n.d., http://www.bioregional.com/flagship-projects/one-planet-communities/bedzed-uk/ (accessed June 7, 2012); BioRegional, "BedZED Seven Years On," 2009, Wallington, Surrey: BioRegional, http://www.bioregional.com/files/publications/BedZED_ seven_years_on.pdf (accessed June 7, 2012).

5   S. Abbaszadeh, L. Zagreus, D. Lehrer, and C. Huizenga, "Occupant Satisfaction with Indoor Environmental Quality in Green Buildings," paper, Berkeley, CA: University of California, June 4, 2006, http://escholarship. org/uc/item/9rf7p4bs (accessed June 7, 2012).

6   Belden Russonello & Stewart LLC, "The 2011 Community Preference Survey: What Americans Are Looking for When Deciding Where to Live," survey conducted for the National Association of Realtors, March 2011, http:// www.brspoll.com/uploads/files/2011%20Community%20Preference%20 Survey.pdf (accessed June 25, 2012).

7   Center for Neighborhood Technology, "The Value of Green Infrastructure: A Guide to Recognizing Its Economic, Environmental and Social Benefits," 2010, http://www.cnt.org/repository/gi-values-guide.pdf (accessed June 7, 2012).

8   See http://www.architecture2030.org/.

9   See http://www.balancedscorecard.org; http://www.globalreporting. org.

10 Jared Diamond, *Collapse: How Societies Choose to Fail or Succeed*, New York: Viking, 2004.

The degree of mitigation will set the course for the degree of adaptation
Image: Vibrant Image Studio/Shutterstock.com

5

Mitigation and Adaptation

Alisdair McGregor

The future is here. It's just not widely distributed yet.

William Gibson

Mitigation strategies reduce carbon dioxide equivalent (CO_2e) emissions with the aim of avoiding the worst consequences of climate change. Adaptation recognizes that some global warming and climate change is inevitable and provides design strategies and policies that are robust to future climate change. The combination of mitigation and adaptation strategies as a means of limiting atmospheric CO_2e concentration at 450 parts per million (ppm) is illustrated in Figure 5.1.

Mitigation and adaptation are complementary; many of the strategies inform each other, and the use of one set of strategies helps minimize the need for the other set. This book has split mitigation and adaptation into separate sections because the type of work that needs to be done for each is managed and controlled differently. Inevitably, there is a question of how to divide resources between mitigation and adaptation projects. The World Bank and development organizations tend to think of the two sets of issues as separate, which is unfortunate. Some, such as James Lovelock, have argued that inertia by governments and big business will stall meaningful mitigation and that we should focus on adaptation.[1] While we do not agree with such a fatalistic outlook, Lovelock's views should be a wake-up call that there is no time to lose in adopting meaningful adaptation measures.

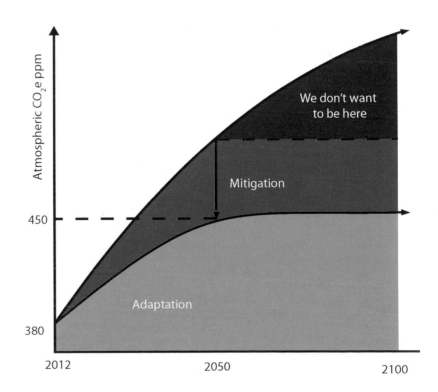

5.1 Mitigation strives to keep CO_2e to 450 ppm, and adaptation is required to deal with impacts of 450 ppm
Graphic: Alisdair McGregor

MITIGATION

Chapter 3 reviewed how different countries have adopted varying strategies to mitigate the impacts of climate change. Members of the European Union are probably the most aggressive in attempting to drive carbon reduction to a point where the worst impacts of climate change can be avoided. Much will depend on whether China and the United States can change trajectory within the time frame predicted by scientists. While mitigation policies and regulations are to be applauded, real reductions are required—the major part of this book is focused on practical approaches to reducing the carbon footprint of the built environment.

Designing buildings or communities to meet or exceed policy targets will not be sufficient, because so often the actual buildings or communities fail to meet their carbon-reduction targets in extended operation. There is a pressing need to develop robust feedback and monitoring so reductions continue to happen and lessons learned from each project are disseminated. With this approach, future projects can be even better. Real-time user feedback will encourage low-carbon lifestyles.

Policy is good and necessary, but it is better to persuade people that mitigation and adaptation strategies are better than the status quo. Trying to persuade people to do something now that will avoid bad things happening in thrity years or more is not easy. Like diets or schemes to save money for college, many are doomed to failure. It is better to demonstrate the short-term benefits or value. Some of the aggressive climate change mitigation policies enacted in Europe were promoted not as something painful to be endured for future benefit but as a near-term strategy to increase local economy and reduce the reliance on out-of-region energy supplies.

In the following chapters, there will be a focus on the direct benefits of mitigation strategies in buildings and communities. The goal is always to make the sustainable solution the most desirable and economic option.

In 2004, Stephen Pacala and Robert Socolow developed the strategy of stabilization wedges as a way of breaking down the seemingly overwhelming task of mitigating the rise in atmospheric CO_2.[2] Each wedge is based on existing technology that can be scaled up and deployed over a fifty-year time frame so that CO_2 could be stabilized by the midcentury at 450 to 550 ppm, or twice preindustrial levels. This requires deploying policy and technology to reduce carbon emissions by 7 gigatons of CO_2 (GtC) per year by midcentury from the business-as-usual case to keep emissions at the 2004 level. Pacala and Socolow proposed splitting the reduction into seven wedges, with each one representing an activity that would start at zero reduction and increase over fifty years to 1 GtC per year by 2054. The major activities are as follows:

- Energy efficiency and conservation
- Fuel switch
- Nuclear fission
- Forest and soils
- Carbon capture and storage
- Renewable electricity and fuels.

Within these six activities, Pacala and Socolow identified fifteen possible wedges. This may give a false impression of how easy it may be to achieve the reduction of 7 GtC per year by 2054, as only seven of the fifteen 1-GtC-per-year wedges are needed. As Socolow states, the wedges "decomposed a heroic challenge into a limited set of monumental tasks."[3] More than seven years after serious work should have started on the wedges, little progress has been made (Figure 5.2).

The nuclear fission option requires fourteen 1-gigawatt nuclear plants to be added globally each year. The Japanese nuclear plant failure following the tsunami in 2011 has all but brought nuclear plant construction to a halt.

Carbon capture and storage has not yet scaled up to significantly reduce CO_2 emissions. Although there is still time for this technology to reach maturity, it will require massive effort to deploy in sufficient quantity to have a major impact. Carbon capture and storage would have a huge effect if it could be universally deployed in the predominantly coal-based power industry of China.

Biomass fuels as a major substitute for fossil fuels have yet to find a technology that is economic or scalable, or, in the case of corn-based ethanol, that even makes sense from a carbon-reduction standpoint. Serious investment in research and technology may result in practical replacement fuels, but it is hard to see this happening before 2020.

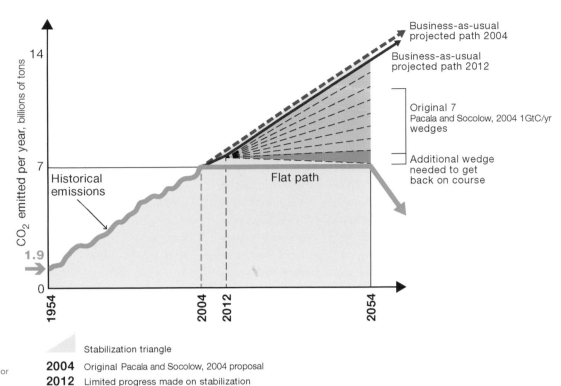

5.2 The stabilization triangle
from Pacala and Socolow, 2004,
adapted for 2012
Graphic: adapted by Alisdair McGregor
from Pacala and Socolow, 2004[i]

△ Stabilization triangle

2004 Original Pacala and Socolow, 2004 proposal
2012 Limited progress made on stabilization

The result of little or no progress on some wedges means that the pressure on energy efficiency and conservation and renewable electricity activities is greater now than when the wedges were first proposed in 2004. Pacala and Socolow (2004)[4] identified two potential wedges in energy efficiency that are directly connected to the built environment:

- Reduce average car miles traveled by half from today by better town planning, improved public transit, and telecommuting/communication technology.
- Cut CO_2 emissions for buildings and appliances by 25 percent by 2054.

The mitigation section of this book focuses on how these two wedges can be achieved or exceeded.

The other wedges that seem plausible include increasing global wind power by 50 times and increasing photovoltaic capacity by 700 times the installed capacity of 2004. These increases seem large, but improvements in technology and falling production costs make these two wedges possible. Improving average car fuel economy to 60 miles per gallon also seems within current technology progression. In 2004, plug-in hybrids were not available and the return of the electric car was still wishful thinking. To accelerate the decarbonization of vehicle transport, we need government policy and incentives.

James H. Williams et al. came to similar conclusions in a study of the technology path needed to achieve the deep carbon reductions needed by 2050.[5] Their study focused on California's goal of an 80 percent reduction below 1990 levels of CO_2 emissions. This target is consistent with the goal of limiting CO_2e concentration to 450 ppm. They concluded that energy efficiency and decarbonization of the electricity supply alone were not sufficient to meet the goals. Electrification of transport and industry was also required. Specifically, they assessed that 75 percent of gasoline fuel vehicles would be replaced by plug-in hybrid and electric-only vehicles. These strategies are viable but will require intense efforts to bring a number of technologies up to scale and to commercial viability. The effort will require coordination of technology progression—for instance, the grid has to be largely decarbonized before a major switch to electric-powered transport.

ADAPTATION

With so much uncertainty in the next thrity years, it would be foolhardy to proceed without a good Plan B. Adaptation is Plan B. However, a good adaptation strategy must never be seen as a reason to delay or eliminate strong mitigation strategies. In other words, adaptation must never become Plan A. Even if all mitigation policies are fully enacted and projects perform as designed, there will still be climate change, so plans for what needs to be done now and what might need to be done in the future require serious thinking.

The location of a project has an impact on the adaptation strategy—low-lying coastal communities are one obvious example. Preparation for different types of change is discussed further in Chapters 17 through 20.

Predicting changes in climate patterns at a regional scale is still a somewhat uncertain science. Scenario planning is a good way of reviewing the possible impacts. Regional trends do emerge, however, and scientists are ever improving their capacity to predict the change we can expect to experience.

The Intergovernmental Panel on Climate Change is the world's most respected scientific body researching climate change science, impacts, and adaptation. Its most recent report—Assessment Report 4 (or AR4)—was launched in 2007.[6] Below are some of the global trends it predicts:

Water
- Increased water availability in moist tropics and high latitudes
- Decreased water availability in mid-latitudes and semiarid low latitudes
- Hundreds of millions of people exposed to increased water stress

Ecosystems
- Increasing species range shifts and wildfire risks
- Ecosystem changes
- Significant increased risk of species extinctions around the globe

Food
- Complex, localized negative impacts on smallholders, subsistence farmers, and fisheries
- Tendencies for decreased cereal productivity; some areas might experience increases in productivity but are at risk for decreases as atmospheric greenhouse gas concentrations continue to rise

Coasts
- Increased damage from floods and storms
- Significant loss of global coastal wetlands
- Millions more people experiencing coastal flooding each year

Health
- Increasing burden from malnutrition, diarrheal, cardiorespiratory, and infectious diseases
- Increased morbidity and mortality from heat waves, floods, and droughts
- Changed distribution of some disease vectors

Conflict and security
- Reduced security of supply of natural resources, e.g., oil, water, food
- Increased pressure of displaced people
- Continued wars launched over oil, possibly wars over water

Issues to consider include the following:

- Changes in health—heat stroke, areas in which diseases are prevalent will move
- Food supply—changes in rain patterns, river flows impacting fisheries
- Sea level rise and storm surges
- Increase in extreme weather events
- Drought
- Desertification
- Benefits to colder climates—longer growing seasons

As the second decade of this century progresses, extreme weather events are already occurring with greater frequency. The financial losses, both direct losses and insurable losses, are enormous. Repackaging adaptation strategies into risk-reduction strategies allows an insurance-based value to be balanced against the up-front cost of the strategies.

The design of buildings is also impacted by a changing climate that can result in buildings becoming more vulnerable and ultimately not meeting their intended purpose. Some examples of climate-related changes and vulnerabilities related to buildings include the following:

- Increased average and extreme summer temperatures; high solar irradiance; coincidence of high wet and dry bulb temperatures causing pressure on mechanical systems; overheating risk; and heating, ventilation, and air-conditioning energy consumption
- Sea level rise, increased rainfall, and soil moisture increasing coastal, fluvial (river), pluvial (drainage), and groundwater flood risks
- Changes in seasonal precipitation and drought risk resulting in water scarcity
- Extreme storms increasing risk of structural failure
- Soil moisture and rainfall causing slope instabilities, subsidence, and heave
- Increased sunshine hours, heat waves, and wind speed and direction changes resulting in external thermal discomfort (creating microclimates and urban heat islands)
- Temperature, changes in seasonal precipitation, and drought risk; changed timing of the seasons increasing risk to green landscape and biodiversity.

Designing with nature is another key component of a sustainable design strategy. We will see in the "Mitigation" section how maximizing the passive system and working with local topography and climate can reduce energy demand, often for a reduced capital cost. Designing for local climate and working with nature was the way all buildings were designed until the industrial revolution. During the last century, designers were able to use technology to overcome nature. We were able to put glass boxes into desert climates by adding more air-conditioning. We could spread cities out farther by adding more freeways and letting people drive from ever bigger houses to their place of work. Unfortunately, it now seems that our efforts to control nature have misfired on a grand scale. Nature has fought back with a vengeance.

The process of designing with nature can be explored further in *Biomimicry* by Janine M. Benyus.[7] The importance of connection to nature has been explored by Stephen R. Kellert and Edward O. Wilson.[8]

In both the "Mitigation" and the "Adaptation" sections, we explore ways of designing with the climate. Adaptation requires that we design for what the climate might be in the future. An essential first step of the design process is to understand the local climate and what it means in terms of the built environment.

NOTES

1 Decca Aitkenhead, "Enjoy Life While You Can," *The Guardian*, March 1, 2008, http://www.guardian.co.uk/theguardian/2008/mar/01/scienceof climatechange.climatechange (accessed June 7, 2012); James Lovelock, *The Revenge of Gaia*, New York: Basic Books, 2006.

2 "Changing Science," *The Economist*, December 8, 2005, http://www. economist.com/node/5278250?story_id=5278250 (accessed June 7, 2012); Stephen Pacala and Robert Socolow, "Stabilization Wedges: Solving the Climate Problem for the Next 50 Years with Current Technologies," *Science* 305(5686), August 13, 2004: 968–72.

3 Robert Socolow, "Wedges Reaffirmed," *Bulletin of Atomic Scientists*, September 27, 2011, http://www.thebulletin.org/web-edition/features/ wedges-reaffirmed (accessed June 7, 2012).

4 See note 2 above.

5 James H. Williams, Andrew DeBenedictis, Rebecca Ghanadan, Amber Mahone, Jack Moore, William R. Morrow III, Snuller Price, and Margaret S. Torn, "The Technology Path to Deep Greenhouse Gas Emissions Cuts by 2050: The Pivotal Role of Electricity," *Science* 335(6064), January 6, 2012: 53–9.

6 M. L. Parry, O. F. Canziani, J. P. Palutikof, P. J. van der Linden, and C. E. Hanson (eds), *Contribution of Working Group II to the Fourth Assessment Report of the Intergovernmental Panel on Climate Change, 2007*, New York: Cambridge University Press, 2007.

7 Janine Benyus, *Biomimicry: Innovation Inspired by Nature*, New York: HarperCollins, 2002.

8 Stephen R. Kellert and Edward O. Wilson, *The Biophilia Hypothesis*, Washington, DC: Island Press, 1993.

i See note ii in Chapter 1.

PART 2

Mitigation Strategies

To maximize wind as a renewable energy source,
the power needs to reach the cities

6

Approaches to Zero Energy and Carbon

Alisdair McGregor

High achievement always takes place in the framework of high expectation.

Charles F. Kettering (American engineer)

This chapter introduces a simple approach to reaching low energy and low carbon buildings and communities. At this stage we are not concerning ourselves with definitions of net-zero energy or carbon. These definitions will follow in Chapters 7 and 8. The following approaches are universal and can be applied to projects of all sizes and varying complexity. They are comprehensive—inclusive of all possible strategies, prioritized—cost effective, and able to achieve net-zero energy and net-zero carbon levels of performance for any building or community. In later chapters, where case studies are reviewed, we'll see how each approach has been applied in practice. Figures 6.1 and 6.3 illustrate the approaches as a sequence of steps. The order of the six steps is

important—too often, design teams jump ahead to working on efficient systems and renewable energy, and then spend much effort and cost on these systems to compensate for missed opportunities in the initial steps.

APPROACHING ZERO ENERGY

The energy approach is most applicable to buildings. Chapter 7 goes into detail about net-zero buildings. What follows is a brief introduction to the route to zero energy.

1. Load Reduction

Before embarking on any energy system designs, it is essential to rigorously reduce the loads the building systems deal with. This means reducing the loads generated inside the building and the external loads entering the building.

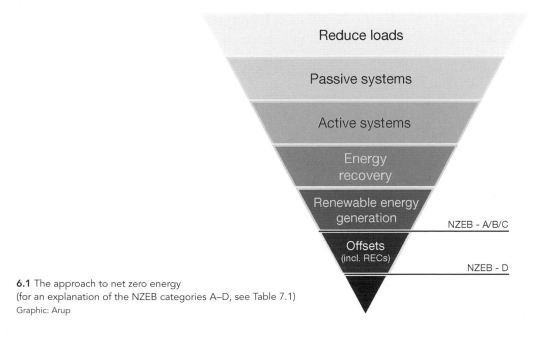

6.1 The approach to net zero energy
(for an explanation of the NZEB categories A–D, see Table 7.1)
Graphic: Arup

The facade, or envelope, is critical in keeping heating and cooling loads down. Performance requirements for the envelope should be set before the architectural design begins. Each facade should be designed for its specific orientation. Ideally the building should be placed on the site to minimize excessive solar gain on the east and west facades.

In many buildings, especially residential buildings, air leakage through the envelope is one of the largest loads, after conduction and solar gain through glazing. Therefore, making sure that the envelope is properly sealed and ventilation controlled is a key strategy for energy reduction.

Internal loads from electrical equipment are often ignored as designers believe that they have no control over these "unregulated loads." But teams can also review internal loads with the users rather than accepting blanket computing and plug loads, which are often overly conservative.[1] If available, measured plug loads from similar buildings can be used for design. Control systems and organizational policies that encourage the shutting down of equipment at night or when people are away from their desks can produce significant savings.

Placing high-heat-gain equipment into specific areas served by dedicated cooling systems reduces the load on general areas. If a building is designed without air-conditioning, a strict load limit has to be set. A rule of thumb is that combined equipment, lighting, people, and solar loads should not exceed 4 watts per square foot. Most of these strategies reduce first cost as well as operating cost.

2. Passive Systems

Once the building loads have been reduced, the design team needs to consider how the building can work with the local climate to provide occupant comfort without using mechanical and electrical systems. Passive solar is a viable design strategy for smaller buildings in a cold climate.

Natural ventilation is a feasible solution for many parts of the world.[2] Even where function and climate prevent completely naturally ventilated buildings, parts of the building or parts of the year may be suitable. Mixed-mode or hybrid options where air-conditioning is only used when really necessary should also be considered. Natural ventilation places restrictions on the design and massing of a building

and demands a fully integrated design approach. Using a decision chart such as the one in Figure 6.2 can help guide the design. Studies have shown that if occupants have control of the operable windows, they will accept a greater range of indoor temperatures.[3]

Daylighting also falls into the passive step. Daylighting is a natural partner to natural ventilation as both require narrow plan buildings and a careful design of the envelope.

3. Active Systems

If the first two steps have been followed, the active systems will be dealing with much-reduced loads. This means that the selection of the active systems should start only once the first two steps have been optimized. The most appropriate systems, sized for the optimized building, can then be developed. Chapter 7 and case studies in later chapters explore the selection of active systems in greater detail.

4. Energy Recovery

Energy recovery is not just about placing heat recovery devices in the exhaust air; it is about a systematic review of the energy paths through the building, looking for opportunities to extract as much use as possible out of every unit of energy that enters the building. There are less obvious opportunities, such as using the return chilled water for radiant cooling where a higher chilled water supply temperature is required. Similar approaches with heating water can make use of cascading temperature demands of different systems. Heat can be extracted from sewer systems, otherwise known as "sewer mining." In mixed-use buildings or developments where some parts need heat while other parts reject it, condenser water loops and heat pumps can move rejected heat to areas that need heat.

5. Renewable Energy Generation

The cost of renewable energy systems is reducing, but careful design is still required to make the economics viable. Photovoltaics are versatile in that they are scalable and can be integrated into the building design. They can also serve multiple functions, such as providing shade (see Chapter 10), which can help reduce the net capital cost.

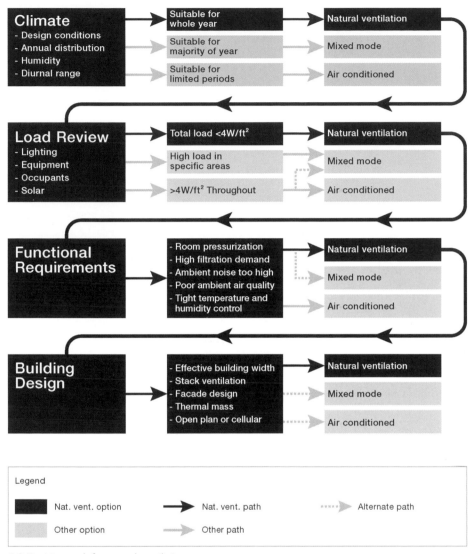

Climate
- Design conditions
- Annual distribution
- Humidity
- Diurnal range

Suitable for whole year	Natural ventilation
Suitable for majority of year	Mixed mode
Suitable for limited periods	Air conditioned

Load Review
- Lighting
- Equipment
- Occupants
- Solar

Total load <4W/ft²	Natural ventilation
High load in specific areas	Mixed mode
>4W/ft² Throughout	Air conditioned

Functional Requirements

- Room pressurization - High filtration demand - Ambient noise too high - Poor ambient air quality - Tight temperature and humidity control	Natural ventilation
	Mixed mode
	Air conditioned

Building Design

- Effective building width - Stack ventilation - Facade design - Thermal mass - Open plan or cellular	Natural ventilation
	Mixed mode
	Air conditioned

Legend

| ■ | Nat. vent. option | → | Nat. vent. path | ⇢ | Alternate path |
| ■ | Other option | → | Other path | | |

6.2 Decision path for natural ventilation
Graphic: Arup

A decision needs to be made at an early stage whether to design the on-site renewable energy systems to supplement energy demand and never export energy or to use the renewable systems to meet a net-zero energy goal on-site. In the latter case, energy will need to be exported to the grid.

Wind power on buildings should be treated with caution. Cost per kilowatt reduces significantly with the size of the turbine, and big turbines don't mix well with buildings. There are some cases where building-mounted turbines make economic sense, but they are uncommon.

Solar hot water is often overlooked but should be considered in climates with significant heating demands or domestic hot water needs.

6. Offset

Offsetting through utility-sponsored green power purchases, direct access utility agreements, and certified Renewable Energy Certificates are the final step to be used to achieve a net-zero energy building (or net-zero carbon building). The authors recommend that Step 6 be limited to compensate less than 20 percent of the energy consumption.

The energy approach can be translated to an emissions basis, as discussed in Chapter 7, and thereby become a pathway for achieving net-zero carbon buildings.

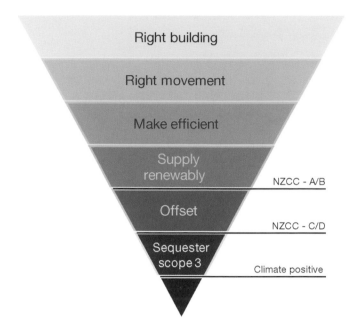

Right building

Right movement

Make efficient

Supply renewably

NZCC - A/B

Offset

NZCC - C/D

Sequester scope 3

Climate positive

6.3 The approach to net zero carbon
(for an explanation of the NZCC categories A–D, see Table 8.1)
Graphic: Arup

APPROACHING ZERO CARBON

The approach to reducing carbon is similar to energy reduction, but carbon reduction requires a more expansive consideration of the impacts of a project and how they are accounted. The carbon diagram is better suited to community or multibuilding developments, but it can still be used for individual buildings. Chapter 8 explores carbon reduction for communities in greater detail.

1. Right Building

There are a variety of options to reducing building. First and foremost, consider just how much building area is really needed. Is a new building or development needed, or can existing buildings be repurposed? Efficient structural design can reduce the amount of steel and concrete used in a building. Can parking structures serve commercial customers by day and entertainment customers by night? Careful selection of construction materials can reduce the embodied energy of the project.

2. Right Movement

Transport to and from a building is often the largest component of its overall carbon footprint. Developments that connect into public transit networks and encourage the use of cycling and walking should always be considered. These requirements are built into most green rating systems for good reason.

3. Make Efficient

Efficiency is covered by Steps 1 through 4 of the energy diagram (Figure 6.1). However, in the carbon-reduction steps, the efficiency of transport also needs to be considered. Hybrid automobiles, car-pooling, shuttle buses, and mass transit are options to consider. For example, as part of Walmart's drive to reduce its carbon footprint, the company improved the efficiency of its trucking fleet by 10 percent (see Chapter 14).

4. Supply Renewably

The use of renewable energy to reduce carbon parallels Step 5 of the energy diagram, but renewable fuel sources for transport are included.

5. Offsets

Carbon offsets are similar to Step 6 of the energy diagram. Offsets, including certified carbon offsets, should be limited and used as the last step in reaching carbon neutrality.

6. Sequester and Scope 3

This may be achieved by markedly restoring the local ecology and sequestration of carbon in biomass and/or by addressing additional Scope 3 emissions inventories (i.e., consumption inventories and the Scope 1 and 2 emissions of others; see box on "Counting Carbon" in Chapter 3). A consensus definition is offered in Chapter 8 for climate-positive achievement. Until widely adopted, care should be taken that carbon credit is correctly documented and is verifiable.

NOTES

1 Lawrence Berkeley National Laboratory (LBNL) Environmental Energy Technologies Division (EETD), "Space Heaters, Computers, Cell Phone Chargers: How Plugged In Are Commercial Buildings?" Berkeley, CA: LBNL EETD, February 2007; J. A. Roberson, C. A. Webber, M. C. McWhinney, R. E. Brown, M. J. Pinckard, and J. F. Busch, "After-hours Power Status of Office Equipment and Energy Use of Miscellaneous Plug-Load Equipment," Berkeley, CA: LBNL EETD, May 2004; C. Sabo, E. Titus, W. Blake, and U. Bhattacharjee, "Best Practices and the Benefits of Delivering Plug-Load Energy Efficiency in Businesses!" http://newbuildings.org/sites/default/files/Plug%20Loads_Sabo.pdf (accessed June 25, 2012).

2 CIBSE, Applications Manual AM 10, *Natural Ventilation in Non-Domestic Buildings*, 2005; J. Parker and A. Teekaram, *BSRIA Wind-Driven Natural Ventilation Systems*, 2005.

3 G. Brager and R. de Dear, "Climate, Comfort and Natural Ventilation: A New Adaptive Comfort Standard for ASHRAE Standard 55," in *Moving Thermal Comfort Standards into the 21st Century: Proceedings of the 2nd Windsor Conference on Thermal Comfort*, Oxford: Oxford Brookes University, 2001; G. Brager and R. J. de Dear, "A Standard for Natural Ventilation," *ASHRAE Journal*, October 2000.

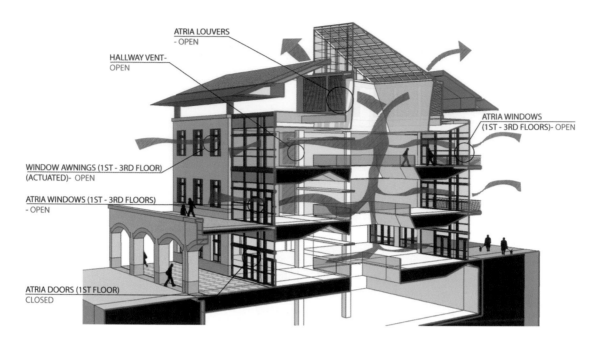

ATRIA LOUVERS
- OPEN

HALLWAY VENT-
OPEN

ATRIA WINDOWS
(1ST - 3RD FLOORS)- OPEN

WINDOW AWNINGS (1ST - 3RD FLOOR)
(ACTUATED)- OPEN

ATRIA WINDOWS (1ST - 3RD FLOORS)
- OPEN

ATRIA DOORS (1ST FLOOR)
CLOSED

COOLING < 82 F OAT

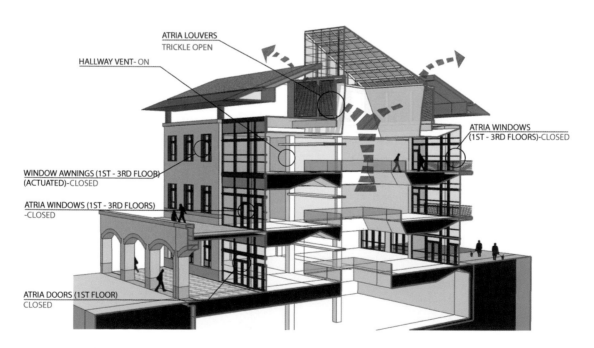

ATRIA LOUVERS
TRICKLE OPEN

HALLWAY VENT- ON

ATRIA WINDOWS
(1ST - 3RD FLOORS)-CLOSED

WINDOW AWNINGS (1ST - 3RD FLOOR)
(ACTUATED)-CLOSED

ATRIA WINDOWS (1ST - 3RD FLOORS)
-CLOSED

ATRIA DOORS (1ST FLOOR)
CLOSED

COOLING > 82 F OAT

The atria of the Jerry Yang and Akiko Yamazaki Environment and Energy
Building at Stanford University are the lungs of the building, passively breathing
the outdoor air and oxygenating the interior
Image: Arup/Boora

7

GUIDELINE CHAPTER

Low-Carbon and Zero-Carbon Buildings

Fiona Cousins

In most people's vocabularies, *design* means *veneer*. It's interior decorating. It's the fabric of the curtains or the sofa. But to me, nothing could be further from the meaning of design. Design is the fundamental soul of a human-made creation that ends up expressing itself in successive outer layers of the product or service.

Steve Jobs

As shown in Chapter 6, the first step in achieving a low- or zero-carbon building is to design and operate a low- or zero-energy building. This chapter will discuss how *zero* is defined and then look at the possible technical solutions for moving closer to zero. There are a great many of these technical solutions—and no single technical or technological move that will guarantee a zero building. Rather, for every building, many separate and complementary solutions will be needed, each contributing a small reduction in overall energy use or carbon emissions.

WHAT IS A ZERO-CARBON BUILDING?

Where Is the Boundary Drawn?

Firstly, we need to define what *zero energy* and *zero carbon* actually mean. The answer depends on the boundary within which you are trying to achieve them.

At the micro scale, imagine that you are sitting at your desk and the boundary within which you are trying to achieve zero energy stretches to your office or cubicle walls and from ceiling to floor. Energy is used by the lights, the computer and monitor, and maybe the heating, cooling, and ventilation systems. For a short period, you might be able to achieve zero energy transfer across the boundary of your office by installing a stationary bicycle and pedaling hard enough to generate the power for all of these systems. Of course, as soon as you become hungry you will need to leave the boundary of your office and find something to eat. Your return from your lunch break will bring a net inflow of energy to the system.

At the opposite extreme, a boundary could be drawn around the world. There are continuous flows of energy to and from the Earth through radiation from the Sun and radiation out to the atmosphere.

These flows balance each other out, which allows the Earth to remain in thermal equilibrium and allows us to say that there is zero net energy across the boundary. Of course, as noted in Part 1, this simple view ignores the change in atmospheric carbon dioxide (CO_2) concentrations, which has changed the thermal properties of Earth's atmosphere and is currently causing a very slight inflow of energy to the Earth.

One implication of this is that a zero energy goal for buildings is not the right goal for the thermal equilibrium of the Earth. A more useful goal is "zero emissions of greenhouse gases that cause global warming," which is usually abbreviated to "zero carbon." Considering carbon rather than energy changes the micro scale example: the metabolic burning of your lunch leads you to emit CO_2, so this temporary zero-energy scenario is not a zero-carbon one.

For the purposes of this chapter, we will draw the boundary at the building perimeter and consider only the energy and emissions associated with fuel burned on-site and electricity used at the site. Chapter 8 describes the energy and emissions issues if the boundary is drawn at community level.

With the boundary drawn at the building perimeter, it is difficult to find a building that can be considered zero energy. Almost all buildings receive sunlight during the day and re-emit the energy as heat overnight. At any given time, energy will be crossing the building boundary. However, unserviced buildings such as barns can be considered zero net energy because the total amount of energy absorbed and the total amount radiated are the same over time. Such buildings can also be considered to be zero carbon for their operations.

Zero Carbon, Carbon Neutral, or Net Zero

Another way to achieve a zero-carbon building is to allow the building to use energy but to provide that energy from sources such as photovoltaics or wind turbines that do not emit CO_2. Renewable technologies are expensive, so the first steps in a zero-carbon design focus on reducing the need for them.

Even after the need for renewables is reduced, the size of a renewable energy harvesting installation is often much larger than the building it serves. Solar and wind energy are also not always readily available when most needed. In a true zero-carbon approach a very

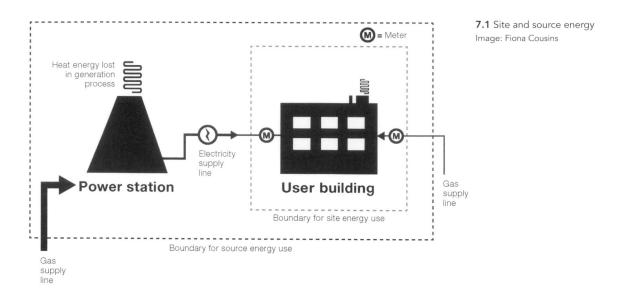

7.1 Site and source energy
Image: Fiona Cousins

A Consensus Definition for Net Zero Energy Buildings

The following criteria have been published by the United States National Renewable Energy Laboratory (NREL)[1] and are intended to align discussion of net zero energy goals.

TABLE 7.1
Net zero energy consensus definitions
Graphic: Arup, adapted from NREL option-based definitions

Net zero energy building definitions		
Site		A *site* NZEB produces at least as much renewable energy as it uses in a year, when accounted for at the site.
Source		A *source* NZEB produces or purchases at least as much renewable energy as it uses in a year, when accounted for at the source.
Cost		In a *cost* NZEB, the amount of money the utility pays the building owner for exported renewable energy is at least equal to the amount the owner pays the utility for imported energy.
Emissions		An *emissions* NZEB produces (or purchases) enough emissions-free renewable energy to offset emissions from all energy used in the building annually.

Renewable energy NZEB options		
On-site supply options	A	Use renewable energy sources available within the building's footprint and directly connected to the building[a]
	B	NZEB A + use renewable energy sources available at the building site[a]
Off-site supply options	C	NZEB B + use renewable energy sources available off-site to generate energy on-site[a]
	D	NZEB C + purchase certified off-site renewable energy soruces[b]

Notes: [a] And connected to the building's electrical and/or heating/cooling networks
[b] And continue to purchase to maintain NZEB status

large renewables system would be installed with energy storage for times when renewable power is not available. An alternative approach is for the building to have both an electrical grid connection and a renewables system. Energy is drawn from the grid when necessary and carbon-free power is sold back to the grid when the renewables system is generating more power than needed. If the energy exported to the grid is equal to the energy imported from the grid, such an arrangement can be described as zero net energy. It might also be described as zero net carbon or carbon neutral if the power exported to the grid allows less fossil fuel to be burned at power stations, thus avoiding some carbon emissions. The time period during which this exchange is measured is usually one year.

It is important to highlight another difference between zero energy and zero carbon. To do this we will use the concepts of site and source energy: if fuel is delivered directly to a building as natural gas or diesel the source energy and the amount of energy crossing the building boundary, or site energy, are about the same. In contrast, when electrical energy is delivered to a building, only a small proportion of the source energy is delivered to the site: typically, fossil-fuel power stations convert only about 40 percent of the source energy to electricity. This means that electrical energy and fuel delivered to the building cannot be directly added together to calculate carbon emissions. To deal with this, different energy use types have different carbon emissions factors that describe the amount of carbon emitted for every unit of site energy—typical values are 0.53 $kgCO_2$/kWh (1.17 $lbCO_2$/kWh) for UK electricity mix and 0.185 $kgCO_2$/kWh (0.41 $lbCO_2$/kWh) for gas use.[2] Carbon emissions correlate much more closely to the source energy than to the site energy.

As we move from building scale to regional scale, transportation energy and its associated carbon emissions, and the energy and carbon associated with other services such as waste management and water provision, also need to be considered. These issues are covered in Chapter 8.

This chapter considers only operational energy use and associated carbon emissions, although a great deal of energy is used for the manufacture, delivery, and installation of materials used in construction. The energy used and the associated carbon emissions are known as *embodied energy* and *embodied carbon*. These concepts are covered in Chapter 2 and further referenced in Chapter 8.

DOWN TO ZERO

To show how a zero energy project might be designed, consider a large mixed-use building in Houston, Texas. The climate provides a heating, ventilation, and air-conditioning (HVAC) challenge: Houston is known for its hot summers, high humidity levels, and small diurnal temperature variations. These factors all make it difficult to design a building that runs without fossil fuel. This building is also large and on a tight site, both factors making it difficult to harvest enough renewable energy or rely on ground-source technologies.

The building is mixed-use and has six parking levels, eight office levels, and twenty residential stories. Using a mixed-use example allows us to look at complementary load profiles within the building—in the real world it is likely that multiple buildings will need to be matched together to find complementary load profiles. Working across multiple buildings needs both cooperation and formal energy agreements. These multibuilding issues are discussed in Chapter 8.

The base building is code-compliant in that it meets the United States energy standard ASHRAE 90.1 2004 and the ventilation standard ASHRAE 62. These standards represent international good practice in terms of the maximum glazing allowances and minimum equipment efficiencies. Amendments to ASHRAE 90.1 continue

7.2 Example building
Image: Arup

to tighten the requirements, but the principles and techniques of energy reductions are the same.

The remainder of the chapter provides details following the energy approach outlined in Chapter 6 (see Figure 6.1).

1. Reduce the Loads

Our example building is similar to many contemporary buildings: while the first step in reducing the thermal loads on a building is to consider its shape and orientation, lettable program and financial return inevitably drive a building's form far more strongly than energy use or the desire for views. In the United States, this typically leads to big floor plates with central circulation cores, as in our example, while in much of Europe there are legal limits on the maximum distance from facade to workspace that make smaller floor plates and offset cores more common. The use of natural ventilation is also more widely accepted in Europe for both residential and commercial buildings. Reducing the peak loads has the important non-energy benefits of improving system control and reducing the size of the installed HVAC systems.

Envelope

The building envelope affects the building load and energy use in four significant ways: heat and coolness are conducted through it; solar radiation passes through windows; air will leak through all joints; and daylight is available near windows. The use of daylight almost always provides an energy benefit, but the solar radiation, conduction, and leakage may be either a bonus or a problem, depending on the time of year and the heat-generating activities within the building. Figure 7.3 shows how these heat transfers can affect the energy use.

Solar radiation on hot days presents the most significant load reduction challenge. It varies hour by hour, peaks at different times on different facades, and can be as much as 50 percent of the total peak cooling load in the perimeter zone. Envelope peak-load reductions will reduce annual energy use but a peak-load reduction of 50 percent may lead to only a 3 or 4 percent annual energy use reduction.

Conduction heat transfer is a smaller component of peak load but it is important to energy use because it occurs whenever the inside

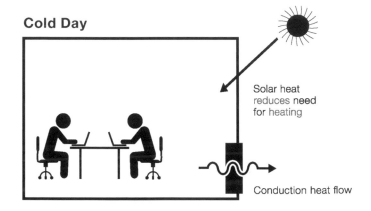

Cold Day

Solar heat reduces **need** for heating

Conduction heat flow

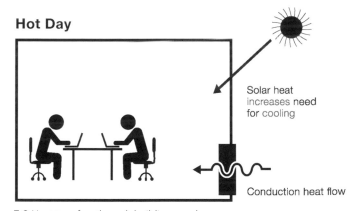

Hot Day

Solar heat increases **need** for cooling

Conduction heat flow

7.3 Heat transfers through building envelope
Image: Fiona Cousins

temperature is different from the outside temperature—almost every hour of the year. Both heating and cooling loads can be reduced by limiting the amount of glazing and insulating the opaque portions of the facade as well as possible. ASHRAE 90.1 and British building regulations set 40 percent glazing as the effective maximum for a code-compliant building.

Air leakage or infiltration also affects energy use because any air that comes in from outside has to be warmed or cooled and potentially humidified or dehumidified to match the internal temperature and humidity. This also happens almost every hour of the year, and the best response is to design a facade that can

Glazing Percentage and Energy Use Reduction

Figure 7.4 shows the effect of glazing percentage on energy use for three cities in the United States and four different glazing types. San Francisco has a moderate climate in both summer and winter, New York has cold winters and hot, humid summers, and Miami has warm winters and hot, humid summers. It is clear that as the glazing percentage rises the annual energy use also rises, whatever the climate. It is also clear that the effect is more marked the poorer the glass performs.

Glazing Type 1 is single glazing, Glazing Type 2 is double glazing, Glazing Type 3 is double glazing with a low-e coating, and Glazing Type 4 is triple glazing with a low-e coating.

Calculations were carried out for a square office building of 100,000 square feet with 10,000-square-foot floor plates. Opaque facade elements were insulated to R-11.

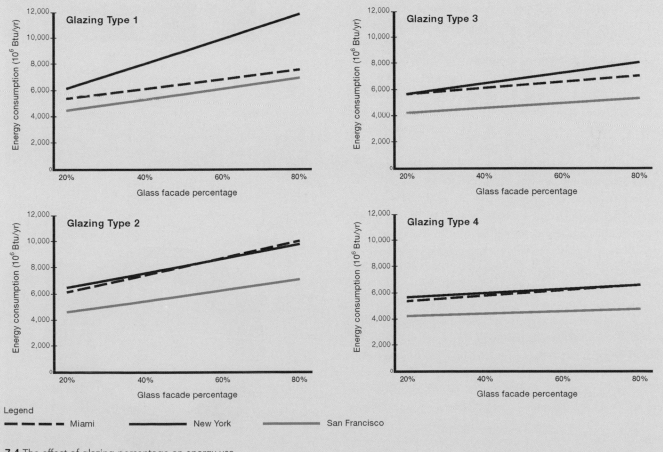

7.4 The effect of glazing percentage on energy use
Graphic: Arup

be properly sealed and to make sure that it is built properly. In large, air-conditioned buildings infiltration is usually controlled by pressurizing the building slightly. In residential buildings, infiltration accounts for a significant portion of the energy use. When buildings are made very airtight care must be taken to place insulation to avoid condensation and ventilation rates must be high enough to avoid mold.

Internal Loads

In considering our example building the first step was to identify the major annual energy uses. This helps set priorities for energy conservation measures. As shown in Figure 7.5 the energy use intensity in the commercial floors is by far the largest and the highest energy use is "other," which includes lights and computers. The residential space has a higher heating load than the office space because there is less equipment and lighting than in the office space; the proportion of cooling and heating is also larger because a bigger proportion of the floor is on the building perimeter. The lower energy use for residential reflects that occupancy and power usage are far denser in an office than in a home and that residential space is assumed to be unoccupied during the day. The garage is not heated or cooled and thus has a low energy-use intensity—the energy used is for lighting and ventilation.

As shown in Figure 7.6, the top three energy uses are lighting, miscellaneous equipment, and cooling. Because the building has a deep floor plate, most of the cooling energy is used to deal with the internal loads and with the fresh air requirements for the occupants. Heating energy is low—as expected for Texas.

Lighting and miscellaneous equipment are two of the components of the cooling load; the other two are the envelope, discussed above, and people. The people load is fixed by the number of people in the office and cannot be changed. The load due to computers and lights can be reduced in two ways: turn off those that are not in use and reduce the amount of power needed to provide the required computing power or light level. Load reductions directly reduce energy use.

Lighting can be turned off manually, which requires occupant training, or through occupancy sensors and daylight controls, which can be implemented electronically.

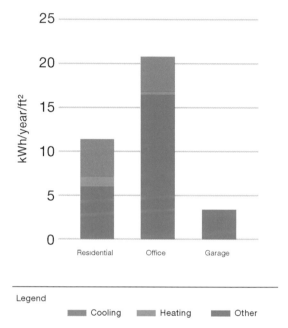

7.5 Comparison of energy use for residential, commercial, and garage space in the example building
Graphic: Nigel Whale

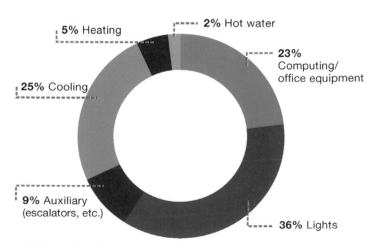

7.6 Distribution of energy use by end use
Graphic: Nigel Whale

The effect of daylight on energy use is small in buildings with deep floor plates, like this example, and the effect in this example's residential spaces is small because the occupied hours are mostly during nighttime. In narrower floor plans there would be a greater effect. Daylight should still be maximized as there is evidence that good lighting can lead to productivity gains.[3]

When used optimally, occupancy sensors are one of the best ways to reduce lighting energy use. Design considerations include the length of time that lights turn on for and the sensitivity and location of sensors. Automatic lighting controls are required by the most recent United States energy codes to make sure that lights are not left on when spaces are not in use.

Lighting is such a major energy user that specific policies have been introduced to reduce energy use: in the United States, the allowable lighting power density for offices reduced from 1.5 watts per square foot to 1.1 watts per square foot between 1999 and 2010. Imports of incandescent lamps, which use a lot of power for the amount of light produced, were banned in Australia in 2009. Lighting power density reductions can be achieved in two main ways: more energy-efficient lamps and different expectations for lighting levels.

Typical office design in the 1990s and 2000s provided uniform lighting levels of 500 lux/50 foot-candles at a desk placed anywhere within the space. Lighting power density can be reduced by providing lower lighting levels throughout the office, where many tasks are simply orientation and circulation. These can then be supplemented by task lighting, specifically where detailed tasks such as reading occur. Bringing the light source closer to the task also improves efficiency and ability for personal adjustment. This change reflects both an increased concern with energy use and changes in working methods, specifically from paper-based to computer-based work.

Computers and other office equipment can be difficult to control automatically because they often take time to start and it is not always easy to tell whether they are in use or not. In spite of this, Energy Star standards for office equipment generally include a requirement for automatic standby. Reducing the power required is therefore the main strategy for reducing computer and office equipment use along with concentrating equipment into data centers.

Office leasing agents have begun to accept lower computer and office equipment capacities reflecting recent reductions in computing power. ASHRAE's recommended design computer load was set at 1 watt per square foot in the mid-1990s but recent research suggests that a lower computer load of 0.25 watts per square foot is adequate for most office spaces.[4] The factors leading to the lower recommendation include more efficient computers, monitors, and printers; greater use of laptops; and the wider adoption of Energy Star standards. There is further potential for load reductions through the use of thin clients and increased use of private and public clouds, both of which centralize computing power. Although some of the energy use is simply displaced from offices to data centers it is often more energy efficient to deal with small areas of high load rather than large areas of low load.

It is important to examine internal and envelope loads first because they create the requirement for building systems, and then load those systems. If the internal and envelope loads can be reduced, then the building systems have to work less hard to maintain comfortable internal conditions.

Changing Design Conditions

The last major method of reducing peak cooling and heating load is to alter the design conditions. Most buildings are designed to hold conditions at 72°F ± 2°F (22°C ± 1°C). Given the right clothing, it is possible for people carrying out light office work to be comfortable between 68°F (20°C) and 78°F (26°C).

Obviously, if a space is heated to only 68°F (20°C) during the winter, then the conduction and infiltration heat loss will be less than if the space is heated to 72°F (22°C). Similarly, if a space is cooled to only 78°F (26°C) in summer, then the infiltration, conduction, and dehumidification loads will be less than if the space is cooled to 72°F (20°C).

Systems are designed for the worst-case condition, so whenever the weather is less extreme, it will be possible to achieve conditions closer to 72°F (20°C). As the most frequent call to maintenance is a hot/cold complaint, it takes a lot of commitment and discipline from occupants and facilities managers to achieve energy savings by altering the heating and cooling set-point temperatures but the potential savings can be up to 15 percent of the annual energy bill.

Wind and Buoyancy Ventilation

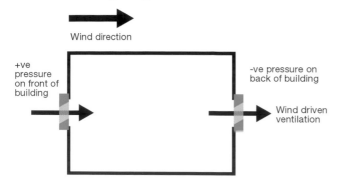

Wind direction

+ve pressure on front of building

-ve pressure on back of building

Wind driven ventilation

a) Cross ventilation

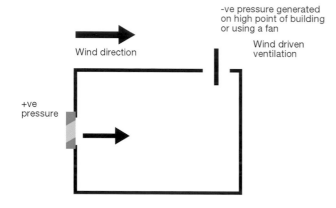

-ve pressure generated on high point of building or using a fan

Wind direction

Wind driven ventilation

+ve pressure

b) Wind-driven ventilation

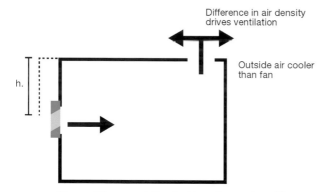

Difference in air density drives ventilation

Outside air cooler than fan

h.

c) Buoyancy driven ventilation

There will be more air flow for bigger temperature differences and bigger heights between lower and upper openings.

7.7 Wind and buoyancy ventilation
Graphic: Fiona Cousins

2. Passive Strategies

Passive strategies are those that need no energy input in order to achieve comfortable conditions within the space. They include natural ventilation, passive solar heating, and passive cooling, and are primarily achieved by manipulating the envelope and thermal mass. Passive strategies can save energy even in air-conditioned buildings if they are applied correctly to the right combination of climate, weather, building function, and base-building system.

Natural Ventilation

Natural ventilation is the practice of providing cooling and ventilation through windows and other openings in the envelope. There are two possible driving forces for ventilation: the wind and buoyancy due to differences in air temperature, as shown in Figure 7.7.

Natural ventilation can offer the advantages of individual control, connection to the outside, and low energy use, when the outdoor conditions are appropriate. Its disadvantages include the possibility of drafty conditions, variable internal temperatures, inability to provide cooling to a temperature lower than the outside air temperature, and no humidity control. Natural ventilation is also not possible in noisy locations or areas with poor air quality.

In the United States, natural ventilation is widely used in residential buildings throughout the country, supplemented by air-conditioning and heating, but is rarely used in commercial buildings. Natural

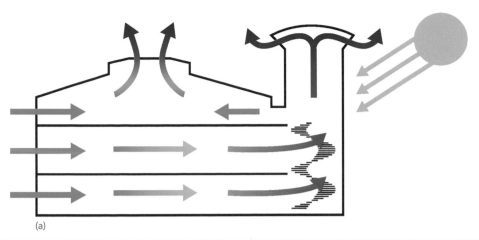

(a)

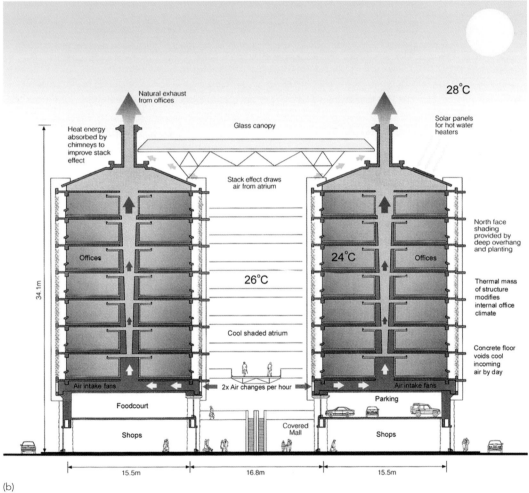

(b)

7.8 Natural ventilation examples: (a) Inland Revenue, United Kingdom,
(b) Eastgate, Zimbabwe
Graphic: Arup

ventilation is more common in Europe, Africa, and Asia. It can be preferred if energy is expensive, outdoor air conditions are usually comfortable, or electricity supply is less reliable.

Passive Solar Heating

In a passive solar heating system, sunshine is allowed to come through a window and fall onto a thermally massive surface that absorbs the heat. The heat is then released over time, reducing the need for heat from other sources. Passive solar heating is most useful in residential buildings in cool climates, where there is little internal heat gain and where heating is regularly needed.

Passive Cooling

There are a variety of forms of passive cooling. One of them involves evaporating water from a pool or pond either in front of or on top of the building. This is generally effective only for the space directly below or next to the pond. An alternative form of passive cooling is to use a rock or earth labyrinth through which air is drawn into the building.

3. Active Strategies

Once the building envelope has been designed to minimize the peak loads, the interior loads have been reduced to the lowest responsible level, and any relevant passive systems have been applied, then it is time to look at the best fit for mechanical and electrical systems. There are strategies that can, and should, be applied in any building. These strategies do not change the overall design of the system, generally have short payback periods, and are primarily based on equipment efficiency and operations. The following equipment can be selected to waste less energy in use: low-loss transformers, premium-efficiency motors, the most efficient version of a chiller, a cooling tower, a fan, pump, and so on. Operational strategies are covered in Chapter 9.

The most appropriate and lowest energy system for any building depends on a number of factors, many of which interact with each other and are very specific to particular buildings. This is not a textbook on HVAC design, so the myriad systems that can be adopted will not be covered in detail. A few general systems and design principles that work to save energy will be described.

As shown in Figure 7.6, the major uses of energy are associated with heating, cooling, and ventilation. HVAC systems generally consist of heating and cooling sources such as boilers and chillers; air and water distribution networks to get the heating and cooling to where it is needed; and terminal units to control the amount of heating, cooling, or air delivered. For simple systems, such as single zone air-conditioning or heating-only systems, the best energy- and carbon-saving approach is to select equipment with the best possible efficiency. This is also the best approach for boilers and chillers. There are four additional areas where it is possible to influence energy use through either system selection or detailed design.

Reducing Prime Mover Energy Use

One of the largest users of energy in every HVAC system is the prime mover (pump or fan) needed to deliver the water or air around the building. The size of the pump or fan is determined by the system pressure drop and the flow rate through the system. The amount of energy used is proportional to the pressure drop and the flow rate through the system. In addition, for any given system, the pressure drop will be proportional to the square of the flow rate through the system. This means that for a given system the energy used is proportional to the cube of the flow rate. Pumps and fans work over a range of pressure/flow combinations but will have greater efficiency at certain operating points.

These simple relationships lead to several means of reducing prime mover energy. First, there will be a range of pumps and fans that can serve any given system, but some will be more efficient than others at the required duty. Prime movers should be selected to operate at their most efficient points, which may mean buying a larger size than the minimum or changing the type of equipment used. As most systems will have variable flow, this can be tricky to do perfectly but should be a design consideration.

Second, if the flow in a system can be reduced at times of partial load, this will reduce the energy used by the prime mover. For example, a drop in flow of 50 percent will result in a reduction in energy use of 87.5 percent, due to the cubic relationship noted above. This is one of the reasons that variable-air-volume (VAV) systems are so common in North America and why variable flow pumping systems are becoming more common.

Third, even if the flow is fixed, for example in a minimum fresh-air system, the system can be designed to minimize the pressure drop. This can be done through the use of larger pipes and ducts or through changing the terminal units. One popular way to do this is through the use of under-floor air systems. These systems tend to have a low pressure drop, because they are usually constant volume and therefore they do not require high pressure-drop terminal units to control flow, they do not use extensive ductwork, and the air tends to move slowly.

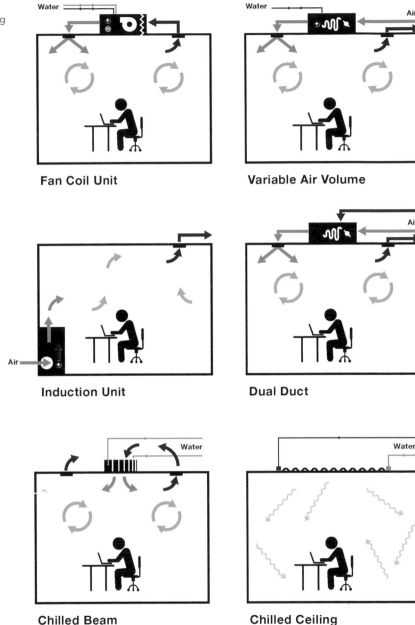

7.9 Different air-based and water-based systems for providing heating and cooling
Graphic: Arup

Using Air-Water Rather than All-Air Systems

In North America, one of the most common HVAC systems in offices is a VAV system. This system distributes heating and cooling using air, controlling the amount of air to each zone using a VAV box. In other parts of the world, pure VAV systems are less common and much of the heating and cooling is distributed using water.

Air is inefficient at moving heat around: imagine carrying a glass of cold water from the kitchen to the farthest point in your home, and compare that with the idea of carrying a glass of cold air over the same distance. You would expect that the water would arrive cool but that the glass of cool air would probably warm up during

the journey, even if you put a lid on it. The reason for this is that air has both a low specific heat capacity, which means that it is unable to transport much heat for every pound of air moved, and a low density, which means that it is hard to gather a pound of air.

In HVAC systems, this means that it is more energy efficient to use water to distribute heating or cooling and to distribute air only for ventilation purposes. It is possible to separate the cooling system from the ventilation system, and it is recommended. There are several established means of doing this—fan coil units, room heat pump units, chilled beams, and radiant floors or ceilings (Figure 7.9)—all of which have different advantages and disadvantages.

Control Strategies to Minimize Energy Use

The third means by which the energy use of mechanical systems can be minimized is through controls. As with lighting and computers, turning systems off when they are not in use is a very effective energy-saving strategy. Mechanical systems are typically sized for a combination of the most extreme weather and the largest likely concurrent internal load. This means that most of the time the system does not need to operate at full output, which means that a control strategy will be used to control the amount of heating and cooling delivered to each zone. A VAV system varies the amount of air delivered to each zone; a fan coil system varies the amount of water provided to the heating and cooling coils.

In addition to these flow reductions, which reduce prime mover energy as described previously, there is often an energy benefit to reducing the amount of fresh air being provided to the spaces. This can be done through time-of-day scheduling but it is more effective to use CO_2 sensors. These detect when the level of CO_2 in a room is more than a certain amount higher than the outdoor CO_2 and control the amount of fresh air to keep this difference at an acceptable level.

Additional control strategies are discussed in Chapter 9.

System Configurations to Minimize Energy Use

Although this is not a book on system design, there are a few system ideas that can save significant energy if they are applied to the right combination of climate and building function. These ideas are not appropriate to every building or every climate, and energy and cost modeling will probably be needed to determine whether they are appropriate.

Evaporative cooling saves energy but increases the use of water. When water is evaporated into an air stream, the air will be become cooler and more humid. This can be useful in two main ways in HVAC systems. First, water can be evaporated into the incoming fresh air, cooling it. This will work only when the incoming air is both hot and dry, for example in desert conditions, and if the air is not then further cooled by the mechanical systems. If the incoming air is humid, then it will not evaporate much water, and if the system cools the air further, then it will also tend to dehumidify the air, so there will be no energy savings. Direct humidification of the incoming air

is appropriate only if the water is very clean and the equipment is very well maintained, otherwise it can lead to Legionnaires' disease or mold growth. Second, water can be evaporated into the exhaust air. The exhaust air in an air-conditioning system will be slightly warmer than and have the same absolute humidity as the room air. In most office systems the exhaust air will easily evaporate moisture and become cooler. The cooled exhaust air can then be passed through a heat exchanger with the incoming fresh air. As discussed in Chapter 2, there is a link between carbon emissions, energy, and water use, and a switch from energy use to water use may result in an increase in associated carbon emissions.

Ground source heat pumps can also reduce energy use. A ground source heat pump system consists of a reversible heat pump—one that can produce both chilled water and heating water—and a means of connecting to the ground thermally. The connection to the ground can be:

- a pair of wells, one of which draws water from the ground and one of which returns water to the ground (an extraction/injection pair),
- a standing column well in which water is drawn from one end of a very tall well and returned to the other, or
- a series of water tubes that are in contact with the ground; these tubes can be embedded in concrete or grout in a mainly vertical arrangement, in which case they are called energy piles, or they can be coils buried over a large horizontal area.

Below about ten feet from the surface, the temperature of the ground is constant throughout the year and tends to remain at about the average air temperature for a given location. In much of Europe, Asia, and the Northern United States, this will be about 55°F (13°C). During the summer, heat from the building is rejected to the ground. This is usually more energy efficient than rejecting heat to the air because the ground is much cooler than the air. This improves the performance of the cooling cycle as well as the heat rejection cycle. During the winter, the cycle is reversed and heat from the ground is provided to the building. In well-designed and well-operated systems, the amount of heat rejected to the ground and the amount of heat drawn from the ground will be approximately equal over the year. A

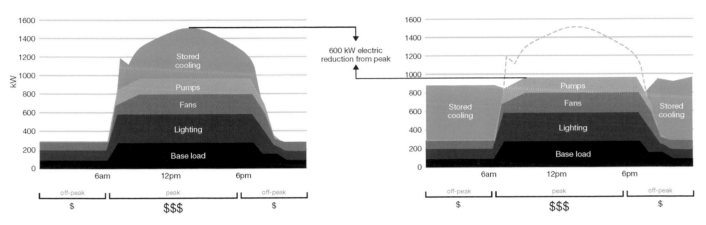

7.10 The effect of ice storage on peak electrical load—note the reduction in peak electrical energy use and subsequent cost saving
Graphic: Arup, Data: CALMAC

significant imbalance will increase or decrease the temperature of the ground over time, which will make the system ineffective. This need for balance limits the locations in which ground source heat pump systems can be applied and may also mean that a supplementary heating or cooling system is needed to deal with the energy that is not part of the balanced load. Applied correctly, this system may reduce the energy use of a building. However, it shifts a portion of the heating energy use from direct fossil fuel use to electric energy, which is often a much more carbon-intensive fuel, and so a careful calculation of carbon emissions is also needed.

Thermal energy storage systems are also often proposed as a means of reducing energy use. The energy can be stored in the building, using the building fabric—this is the system used at The Gap in San Bruno, for example—or, more commonly, in an ice-storage system. The concept for building-fabric storage relies on diurnal outdoor temperature variations, and the climate and building arrangement determine whether it will work. The energy concept for an ice-storage system is that running chillers at night will improve the efficiency of the chillers because the heat rejection cycle can operate at a lower temperature. However, the need to produce ice will decrease the efficiency of the chillers because the chilled water cycle will also have to operate at a lower temperature. There will also be an overall decrease in energy because the ice then has to be stored and converted back to chilled water—all processes that are less than 100 percent efficient. Ice storage systems will save the most energy in climates with a big diurnal variation, because chiller operating improvements will outweigh the other considerations.

Ice storage systems will often save energy costs because they use nighttime electricity to meet daytime cooling loads. Depending on the rate structure used, this will reduce the demand charge, allow peak load management bonuses from the utility, or simply use cheaper nighttime electricity. As with all strategies that change the timing of the use of electricity, there can also be an effect on carbon emissions: in some places the electricity produced from base-load plants (the ones that operate at night) will have lower carbon emissions than the electricity produced from peaking plants (the ones that operate during the day). If this is the case, then ice storage can be a very attractive carbon reduction strategy. It could also be used as an electricity storage mechanism in a smart grid. Refer to Chapter 8 for more details.

4. Energy Recovery

Energy recovery is a fundamental component of active system design. The simplest forms of heat recovery—the use of air-side and water-side economizers—are required by code in the United States and the United Kingdom.

There are many other ways that heat can be recovered in systems. In air systems, heat is frequently recovered from the exhaust air and transferred to the fresh air using a heat exchanger, a run-around coil, or a heat pipe.

There are other, less obvious ways to recover heat: buildings that use heat pumps in each room connected to a common circuit transfer heat from the warm rooms to the cool rooms through the circuit. Any imbalance is made up by a cooling tower and boiler

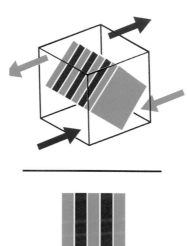

Vertical flat plate
50–70%

(a)

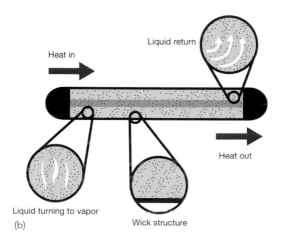

Heat Pipe Operation

Heat in

Liquid return

Heat out

Liquid turning to vapor

(b)

Wick structure

7.11 Energy recovery devices:
(a) air to air heat exchanger, (b) heat pipe
Graphic: Arup

serving the whole building. Condensing boilers work to recover heat from the boiler exhaust gases by passing relative cool water through the flues, extracting heat. Buildings that have excess heat at a high temperature, perhaps from a combined heat and power plant, can use that heat directly for heating, or to drive an adsorption or absorption chiller, or to provide desiccant dehumidification. Air systems can sometimes be designed to divert air that would otherwise be exhausted to rooms where it will be used.

For the example building described at the beginning of this chapter, a number of these strategies and systems were analyzed, including improved lighting, dedicated outdoor air systems, more efficient cooling towers and chillers, additional economizer systems, increased insulation, and under-floor air. The results are presented in Figure 7.12.

Each of the bars on this chart represents a different energy-saving measure, implemented cumulatively. For a building of this size, it is

clear that there is no magic bullet. The only way to reduce the site energy use is to reduce the energy use on all fronts. This includes buying the most efficient equipment possible, minimizing the run time of equipment, and providing heating and cooling in the most energy efficient ways.

All these measures get the building energy use down to about 60 percent of the code-compliant energy use. This represents about the lowest energy use that can be achieved using current technologies and attacking the demand side only.

5. Renewable Energy Generation

Once the energy demand has been reduced to be as low as possible, we can start to look at more carbon-efficient ways of generating electrical power. The most common approaches are to use renewable energy sources such as solar, biomass, and wind, or to use a cogeneration (cogen) plant.

Site Energy with CHP Cases

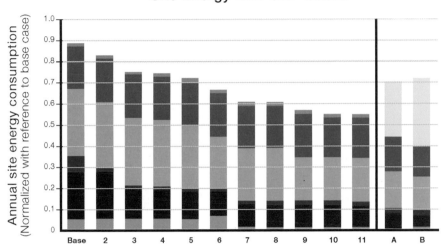

Source Energy with CHP Cases

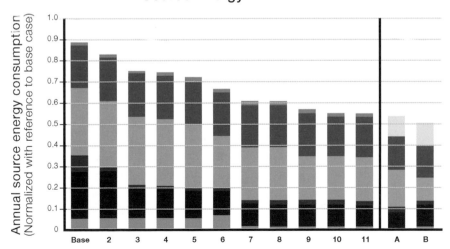

Legend
- Heating
- Cooling
- Auxillary
- Lights
- Miscellaneous equipment
- Domestic hot water
- Combination heat and power (CHP)

7.12 Comparison of energy use with implementation of different mechanical system strategies
Graphic: Nigel Whale

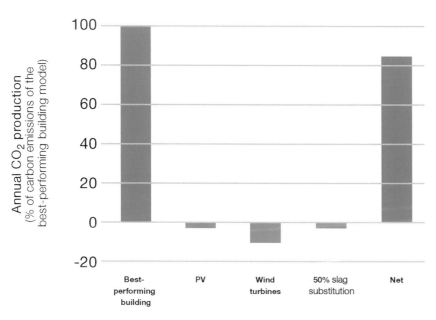

7.13 The effect of adding two 6-kilowatt turbines and covering the roofs of the building with photovoltaics, in terms of carbon
Graphic: Nigel Whale

Cogen is also known as combined heat and power, and it can be implemented at the building scale or at the neighborhood scale. The rules of thermodynamics dictate that when you convert fuel to work there is a lot of heat rejection. The generation of electrical power is a process of converting fuel to work and from there to electrical energy. If the power is generated remote from a heat use, then most of the heat is wasted and is discharged to the atmosphere. If the power is generated close to a heat use, then the heat can be used to do useful work. This might include heating the building, heating domestic hot water, or providing cooling using a steam-driven absorption or turbine-driven chiller. In Texas, the most likely use is for cooling. Fuel cells are also a form of cogen.

The top chart in Figure 7.12 shows the site energy reductions as we move through all the energy-saving cases, then shows a large increase in site energy use when cogen is added. This is as expected, because, by installing cogen at the site, we have shifted fuel use from the electric power station to our site.

However, if we look at source energy, the use of cogen involves an overall reduction in the amount of fossil fuel used and therefore the amount of CO_2 emitted. The lower chart in Figure 7.12 shows the energy savings in terms of source energy, which correlates much more closely to carbon emissions.

At this point, the opportunities for reducing the energy use at the building have been exhausted, so the only way to get to zero carbon is to consider renewable energy sources. In Houston, these include solar and wind energy. In other parts of the country it would be possible to include volcanic or geothermal energy.

Figure 7.13 shows the effect, in terms of carbon, of adding two 600-kilowatt turbines and covering the roofs of the building with photovoltaics. It is clear that the amount of energy that can be provided by building-based renewable sources is tiny for a large building like this, on a small site.

Both cogen and renewable energy generation are often more appropriately provided at the grid or neighborhood scale than at the building scale, because there is frequently a better match between load profiles. Chapter 8 describes in more detail how cogen and renewable sources can be added to the grid and how this can help reduce carbon emissions.

6. Offsets

Once the building energy use has been reduced as far as possible and when the incorporation of renewables at the building cannot meet the remaining energy use, then the building can achieve net-zero status only by purchasing off-site carbon offsets. These may include utility-sponsored green power purchasing programs, "direct access" renewable energy agreements, certified renewable energy certificates, and certified carbon offsets.

There are many ways of providing offsets, discussed further in the "Carbon Offsets" box.

Offsets may be purchased by organizations that need to reduce their carbon emissions to meet either voluntary or mandated carbon emission reduction targets.

Carbon Offsets

Carbon offsets are generated when voluntary reductions are made to greenhouse gas emissions that would otherwise not have been made. Reductions can be made through alternative processing of farm manure, renewable energy generation, methane capture from landfill to be used as fuel, land-use changes or similar. If the reduction is truly additional and can be traced using complete, accurate, comparable, reliable, and transparent reporting standards then it can be sold as an offset.

Offsets are also known as Green Tags, Tradable Renewable Certificates, or Renewable Energy Certificates. It is also possible to generate carbon offsets through energy efficiency measures. Offsets generated this way are sometimes referred to as White Tags.

LOW-CARBON AND ZERO-CARBON BUILDINGS IN SUMMARY

There are many steps to getting to zero energy, and we can take them at many scales. Firstly, occupants should be trained to consider their expectations and motivation. Then we, as designers, should design our buildings to use as little energy as possible and to harvest as much energy as possible on-site. Then we need to cooperate with our neighbors to share equipment and power; and finally we should move toward a regional renewable-power infrastructure, bearing in mind the amount of embodied energy required to construct one.

NOTES

1 Shanti Pless and Paul Torcellini, *Net-Zero Energy Buildings: A Classification System Based on Renewable Energy Supply Options*, NREL/TP-550-44586, Golden, CO: National Renewable Energy Laboratory, June 2010. For source of Table 7.1 see: Nancy Carlisle, Otto Van Geet, and Shanti Pless, "Definition of a Zero Net Energy Community," National Renewable Energy Laboratory, Technical Report, NREL/TP-7A2-46065, 2009.

2 Defra, *BNXS01: Carbon Dioxide Emission Factors for UK Energy Use*, briefing note, version 4, 2009, http://efficient-products.defra.gov.uk/spm/download/document/id/785 (accessed June 26, 2012).

3 Located near Los Angeles, California, VeriFone, Inc. constructed a new daylit Worldwide Distribution Center and reported increased productivity a year and a half after it started using its new building. Productivity increased by more than 5 percent and total product output increased 25 to 28 percent, making the new building more cost-effective than first predicted (W. R. Pape, "At What Cost Health? Low Cost, As It Turns Out," *Inc. Online*, August 8, 1998). In 1983, Lockheed Martin designers successfully increased interaction among the engineers by using an open office layout with integrated daylighting in its offices in Sunnyvale, California. This increase helped boost contract productivity by 15 percent (J. Pierson, "If Sun Shines in, Workers Work Better, Buyers Buy More," *The Wall Street Journal*, November 20, 1995).

4 Christopher Wilkins and Mohammad Hosni, "Plug Load Design Factors," *ASHRAE Journal*, May 2011.

Village Homes community, in Davis, California, originally developed in the 1970s, has become a model low-carbon community where the homes are so desired they sell for a 15 percent premium, often by word of mouth

Image: Ken Kay

8

GUIDELINE CHAPTER

Low-Carbon and Climate-Positive Communities

Cole Roberts with Mark Watts

There is a quality even meaner than outright ugliness or disorder, and this meaner quality is the dishonest mask of pretended order, achieved by ignoring or suppressing the real order that is struggling to exist and be served.

Jane Jacobs, *The Death and Life of Great American Cities*[1]

If we are to spend up to 2 percent of global gross domestic product on climate change, at what scale are our investments most effective?[2] How much should we invest in better buildings and better communities?

More importantly, if we are to transition this expenditure from a politically unpalatable sacrifice to a prudent and enriching investment in our future—one that strengthens the bonds of our communities, builds economies, and creates a more just and healthy world for us all—how do we invest the money wisely?

Engineering expertise shows that the interdependency and economy of scale that occurs beyond the scale of individual buildings

is often where some of the greatest opportunities exist. Take renewable energy as an example. Although there is merit in distribution of renewable energy technologies on individual buildings, renewable energy on individual high-rise buildings in a dense downtown core invariably suffers from challenges ranging from small productive output relative to the dense energy demand of the buildings to limited "fuel" caused by shading, turbulent wind conditions, and small floor plates. By comparison, suburban applications are more likely to have adequate roof area, limited obstructions from adjacent properties, and high productive capacity relative to the energy density of the building. Indeed, much of building renewable energy integration occurs in rural settlements without access to grid-supplied energy sources. This would seem to suggest the greater sustainability of low-density development. However, the benefit of density, in contexts ranging from transportation systems to job creation to health care, points in the opposing direction—each is immensely more effective in well-functioning urban areas (Figure 8.1).

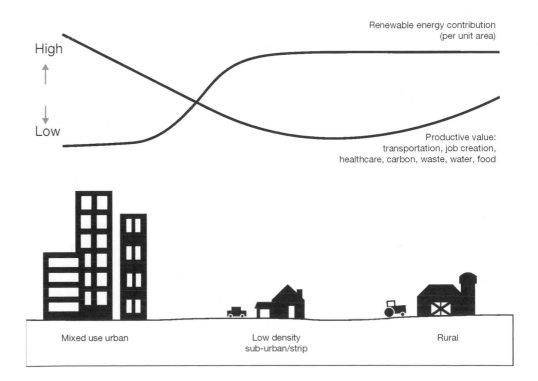

8.1 The productive value of suburban development is low despite the relative renewable energy potential

Graphic: Arup

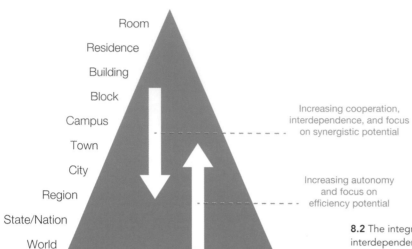

Room
Residence
Building
Block
Campus
Town
City
Region
State/Nation
World

Increasing cooperation, interdependence, and focus on synergistic potential

Increasing autonomy and focus on efficiency potential

8.2 The integration pyramid—relationship of scale to cooperation, interdependency, and synergy
Graphic: Arup

Answering the question of where to invest is no small challenge. It speaks to a host of actions that we take every day in every corner of our society, and it may force a move beyond the boundaries of expertise, the precedents of cooperation, and the legal framework of our codes and international treaties. But it is possible. It is analogous to the transition toward integrated (or holistic) design (see Chapter 10), which has increasingly found a maturity in the building design fields, academia, and medicine, a move away from siloed hierarchical expertise in favor of integrated cooperation under emergent leaders. A similar integrated design and planning effort is needed at scales beyond the individual building.

And if we do agree that the interdependency and economy of scale that occurs beyond the scale of individual buildings is where some of the greatest opportunities exist (see the integration pyramid in Figure 8.2), more questions follow. When does it make sense to imagine systems at the community scale—creating in effect a "network" of buildings? At what scale do select energy, water, and waste technologies make sense? What are the implications of systems optimizing at different scales? What are the variables and tools that support decisions about how and when to proceed? What are the financial and ownership implications?

This chapter makes an attempt to consider each of these questions within the context of climate change mitigation and improved community resilience. It begins with an approach to zero-carbon communities and concludes with a detailed discussion of systems at a scale beyond the individual building.

WHAT IS A ZERO-CARBON COMMUNITY?

Where Is the Boundary Drawn?
As discussed in Chapter 7, the carbon sources included in a net-zero carbon claim are important for transparency of accounting. Unlike net-zero-carbon buildings (NZCBs), net-zero-carbon communities (NZCCs)[3] are complicated by the need to consider additional emissions sources, most notably transportation-related emissions (which include trips that originate and terminate within the community and those that pass through the community), as well as waste emissions from landfills, energy emissions from water services and treatment, and fugitive[4] emissions from a variety of sources. Additional complication results from diverse ownership interests and limited access to data (e.g., in the case of a city), the presence of both new and existing buildings (e.g., in the case of an existing academic or corporate campus), and the extended time horizon of community carbon-reduction strategies (e.g., legislated renewable portfolio standards, low-carbon fuel standards). Despite this complexity, significant progress is being made in defining NZCCs, enforcing their carbon-reduction targets, and establishing repeatable approaches to cost-effective achievement.

The Baseline Inventory
Just as it is important to understand how you spend money when setting a personal or business budget, a first step to creating low-carbon communities is understanding how and where carbon is emitted. Most commonly, this starts with a baseline inventory of

Community Consumption of Goods and Services

Community emissions summaries typically exclude the largest source of community emissions: the indirect emissions that result from the buying of goods and services. These indirect sources can be linked back to the same transportation and building emissions that occur where the products are being manufactured. However, the emissions in those locations (e.g., Chinese industrial cities) are not internalized into the products exported for sale and hence are simply "not counted." Studies have shown that these indirect emissions may be up to 50 percent greater than the total Scope 1 and 2 emissions in a typical community-production-based inventory.[5] Such a realization underlines the importance of policies to internalize climate change costs into goods production (e.g., through carbon import tariffs or a global treaty) and prepare Scope 3 dominant-consumption-based inventories. For the vast majority of goods, the actual cost of embodied carbon is less than 1 percent of the retail cost of the goods—a cost easily absorbed into the global economy as a whole.[6]

emissions sources across various segments of the community, aligned to a Scope 1/2/3 reporting protocol intended to eliminate double counting and assign responsibility.[7]

Although such an inventory is important, look at enough of these baseline inventories and you'll start to recognize commonalities suitable for generalization to communities that may not yet have an inventory completed:

- Transportation and building energy use are often over 80 percent of emissions (over 50 percent and over 30 percent, respectively).
- Lesser sources, but still significant, include waste, water, and fugitive carbon emissions.
- Medium-sized communities that have a higher dependence on cars have relatively higher transportation-related emissions compared to highly localized small communities or large communities with effective transit systems.
- Commercial, residential, and industrial emissions track closely with the amount of each building type that is present in the city. Cities with high industrial development have high industrial emissions, bedroom communities yield high residential emissions, and downtown central business districts yield high commercial emissions.
- Cities in developed countries tend to have higher carbon intensity.

Look closer and there are also notable differences:

- Dense communities have lower carbon intensity per person. Better transit, smaller conditioned spaces, shared low heat transfer walls, and optimized city-scale systems are often contributors.
- Between cities that have comparable quality of life, carbon intensity is not the same (i.e., high quality of life *does not* require high carbon emissions). Examples include Stockholm, Kyoto, Paris, Hong Kong, and Singapore.[8]
- Car-dependent communities at any size have significantly higher per capita emissions.

These observations point to three significant insights: (1) in a resource-constrained global economy, low-carbon communities have a competitive advantage, (2) the greatest *opportunity* for change is actually in high-carbon-emission communities (though the *ability*, *motivation*, and *triggers* discussed in Chapter 11 are needed for action), and (3) current technology exists for significantly reducing carbon intensity.

The Carbon Profile of Communities

In addition to annual emissions, it is often helpful to consider the emissions profile of a typical day or week—the timing of a community's activity will impact the amount of carbon released. There are a few reasons for this:

- *Time-dependent emissions factors* The power plants that provide electricity to communities have differing emissions

The C40:
City Mayors Taking a Lead toward Sustainable Cities
Mark Watts

In spring 2005, as the world's most powerful nations (G20) convened in Scotland, London's mayor, Ken Livingstone, called a parallel meeting of the "C20 cities." Britain's prime minister, Tony Blair, had put climate change on the G20's agenda for the first time, but Livingstone doubted that much would come of it. City governments, however, were increasingly taking up the challenge. Spurred by his deputy mayor, Nicky Gavron, Mayor Livingstone hoped to create an alliance of his peers across the world's great metropolises to demonstrate to national leaders how a low-carbon future could be achieved.

Nearly seven years later the C40, as it became known, is a thriving organization of forty of the world's most renowned large cities, from New York and Toronto to London and Paris, Beijing and Tokyo, Delhi, Bangkok and Sydney, São Paulo and Mexico City, Addis Ababa and Johannesburg. Now under the leadership of Mayor Michael Bloomberg of New York, it has merged with Bill Clinton's Climate Initiative and also incorporates eighteen affiliated smaller leading sustainable cities such as Copenhagen, Stockholm, Portland, San Francisco, and Curitiba.

The importance of the C40 to the task of delivering sustainable cities is threefold:

- Together its members represent nearly 300 million people, 10 percent of global carbon emissions, and 18 percent of the world's gross domestic product—what these cities do affects the world.
- The majority of members have genuinely made tackling climate change a priority: a recent study by Arup found nearly 5,000 separate "actions" had been taken by C40 mayors to cut emissions since 2005.
- The network itself demonstrates the ability of multinational partners to share knowledge and use it effectively. There are clear links between knowledge shared by leading cities in the C40 and subsequent action by others.

For C40 cities in rapidly developing parts of the world, the challenge is often how to build a new, sustainable urban form. Nowhere epitomizes this challenge more than China, which in 2010 accounted for over half of all the world's new building construction (measured in square meters). But equally strikingly, the Ethiopian capital, Addis Ababa, is awash with cranes as a raft of new houses, schools, hospitals, roads, and railways take form. Where previously 80 percent of the population lived in unplanned slums with only basic sanitation, that figure is now down to (a still staggering) 40 percent, and falling.

In postindustrial cities, however, there is a different problem: at least half of today's buildings will still be in use in 2050, when carbon emissions need to be 80 percent lower. The focus is thus on "retrofit": improving thermal insulation, reducing the need for mechanical cooling, and creating "smart" buildings that regulate themselves in the most cost-effective and carbon-efficient manner possible.

Here the cities of Berlin and Toronto have set a lead, the former demonstrating 10 million euros a year savings from achieving a 25 percent average reduction in energy usage in 1,300 municipal and commercial buildings. The energy performance contracting method used to procure this effort, whereby private capital funds capital works and is repaid out of a proportion of future energy bill savings, has been repeated across the C40 network. Toronto's former mayor, David Miller, started an equally ambitious Tower Renewal program to retrofit 1,600 city apartments as part of a broader greenification plan.

Perhaps the most interesting impact of the C40 has been in transport policy, particularly the revival of cycling as a tool of big-city transport policy. A combination of inspiration from affiliate cities such as Amsterdam and Copenhagen, where cycling accounts for over 30 percent of trips, plus the high-profile Paris Vélib' cycle hire scheme has led to many imitators across the C40, from Mexico to London. Perhaps most significantly, this year Shanghai, where cyclists have been displaced by a roaring onrush of car drivers in recent decades, reversed an early decision to ban cycling from its major roads.

Mayor Livingstone's politically brave, and successful, decision to introduce congestion pricing in central London can be seen as an early carbon tax. C40 affiliates Stockholm and Milan have since introduced similar schemes; New York has tried and failed; but Copenhagen and San Francisco may be set to follow.

Arup's joint study with the C40—*Climate Action in Megacities: C40 Cities Baseline and Opportunities*—demonstrated that big city mayors are able to exercise significant powers across most of the sectors where carbon emissions are highest: from buildings to transport to water and waste. The one area where, on average, city mayors are relatively powerless is the generation and distribution of energy itself.

There are some exceptions: the Scandinavian cities have built up highly efficient, city-owned combined heat and power systems since the oil price shocks of the 1970s, and this is one reason they register some of the lowest per-capita carbon emissions.

But among the majority of C40 mayors, the most frequently used method to decarbonize energy supply is to generate energy from waste. This makes sense—over half of C40 members exercise strong powers over waste collection or treatment (or both). At present, there is a focus on landfill gas capture: São Paulo, for example, already meets 7 percent of the city's electricity demand in this way, and every other C40 city in South America now has plans to follow suit. But in the longer term more efficient methods, such as anaerobic digestion and gasification, are likely to be used.

The C40 has also been a forum for innovation, with its electric vehicle group being one of the most vibrant parts of the network's collaboration and widespread focus on smart grids and metering. However, here action has yet to match aspiration—only four member cities report having installed smart meters in their own buildings, for example.

From its foundation, the principle behind the C40 has been simply: every one of the member cities is a world leader in something, but it is very hard to be at the vanguard of everything. By pooling their knowledge, sector by sector, the C40 makes it much easier for mayors to deliver best practice across the board, safe in the knowledge that they have a successful example to imitate, or a failure to learn from.

The proof is in the pudding, and Arup's research shows that sharing knowledge is working—over half of all the low-carbon actions taken by C40 cities were delivered in the last surveyed year (2010–11), with a clear ramping up of activity since the twenty vanguard cities first got together in 2005.

While global carbon emissions have continued to ratchet upward over the same period, progress in the G20 nations has been a little slower.

factors depending on their technology and fuel source. Base-load plants, which are intended to operate continuously with relatively slow changes in output, will release a set amount of carbon per unit of electricity supplied to the grid. Peaking power plants, which have the ability to respond to short duration power fluctuations (or peaks) but may have higher operating costs, typically have another emissions factor. Wind and solar energy plants may have a zero emissions factor; however, they track the availability of wind and sunshine. Many governments and utility regulators publish such time-dependent emissions factors. Where the information isn't available, it is often possible to gauge when the lowest carbon energy sources are available in a community, based on the type of plants providing base-load and peaking supply. Nuclear, natural gas, hydroelectric, and renewable energy plants have the lowest emissions. Oil- and coal-fired plants have the highest emissions.

- *Demand diversity* Coincident activities within a community can reduce efficiency. The daily commute traffic jam, with its frequent breaking and accelerating, is a perfect example. The converse is that mixed activities within a community can enhance efficiency. The otherwise highly variable electrical load of offices and residences between day and night is actually quite flat when combined together in a mixed-use community. Generally, the more such peaks can be flattened out and diversified—whatever the resource flow: electricity, natural gas, water, traffic—the better it is for infrastructure cost-effectiveness and reducing carbon emissions.

Net Zero or Climate Positive?

The best definition of NZCCs is the consensus definition published by the U.S. National Renewable Energy Laboratory (NREL).[9] This definition builds on and is internally consistent with the NZCB definition published by NREL and the U.S. Department of Energy referenced in Chapter 7.[10] The definition offered by NREL for NZCCs is "[a community] that has greatly reduced energy needs through efficiency gains such that the balance of energy for vehicles, thermal, and electrical energy within the community is met by [on-site or off-site] renewable energy." The definition goes on to further "grade" NZCCs from A to D, based on their renewable energy supply (Table 8.1).

TABLE 8.1

NZCC definitions

Graphic: Arup, adapted from NREL option-based definitions[11]

Grade	Criteria
A	100% emissions load met by renewables in the built environment and on unbuildable brownfield sites within the community
B	Any portion of the emissions load met by renewable generation on community greenfield sites or from off-site renewables used on-site (e.g., imported biomass)
C	A and/or B *and* remainder portion of the emissions load offset through RECs that add new grid generation capacity
D	100% of the emissions load offset through RECs that add new grid generation capacity

NREL also defines a low- (or near-zero-) carbon community as "a community that produces at least 75 percent of its required energy through the use of on-site renewable energy." It is assumed that such a high renewable energy percentage would have been preceded by aggressive energy efficiency investments.

There is not yet a consensus definition for a beyond-zero or climate-positive community. Many climate-positive definitions tend to focus on a reframing of the approach to climate change, i.e., a focus on the positive opportunities rather than the negative activities—more good not just less bad. Beyond this reframing, there are working definitions that, though not yet widely adopted, suggest an emergent definition. The first working definition is the focus on sequestering carbon beyond the net-zero achievement to address historical carbon debt (past emissions). Hence an action by an NZCC to restore ecologically diverse, permanently protected, high-carbon-yield forests,[12] sustainable farming practices, wetlands, etc. would justify a climate-positive claim. The second is the focus on addressing emissions beyond the Scope 1 and 2 boundaries, such as community consumer goods and services choices, and sustainable food choices.[13] The rationale is that such an improvement is making goods producers that would not otherwise move toward net-zero carbon do so. Hence, an action by an NZCC that resulted in quantified improvements in low-carbon purchasing behavior could justify a climate-positive claim.

For the purposes of this book, the two definitions are merged into a consensus definition. A net-positive-climate community (NPCC) is a community that achieves NZCC status and reverses historical carbon debt by sequestering carbon at an annually positive rate for the life of the community, and/or measures and reduces Scope 3 emissions using established accounting protocols. This may occur through additional offsetting, ecological restoration activity, and community behavioral shifts.

DOWN TO ZERO AND TOWARD CLIMATE POSITIVE

Once an understanding of the baseline inventory (what it includes and what it doesn't), the time-dependent emission factors in a community, and the nature of the NZCC "grade" are set, there is still a need for a cost-effective and repeatable approach to get to zero—or near zero. Such an approach must be comprehensive, so that it includes all potential technologies, policies, and actions. It must also be prioritized, so that it makes sense from one step to the next and achieves a cost-effectiveness in totality that is repeatable from one implementation to the next. The approach outlined in Chapter 6 and detailed below achieves exactly this.

Before we get to the detailed approach itself, it is helpful to explain what is meant by "makes sense from one step to the next." Although the steps are generally sequenced with the lowest-cost solutions first (e.g., efficiency is generally cheaper than new renewable supplies), they are also sequenced because of their interrelationship to one another. To understand this, consider the systems interrelationship of a watershed where people and ecology upstream and downstream regularly impact each other for better and worse.

The Low-Carbon Community Shed

The same upstream and downstream connectedness that holds true for a watershed is also true of all our resource systems. You can feel this in our commute sheds tangibly when you're bound in a traffic jam or standing in a packed bus with others on the way to and from work. It is also true of our food sheds (think health and farmland preservation), our energy sheds (think emissions and time-of-use billing), our local-economy sheds (think "buy local"), and our carbon sheds (which are indexed to energy, water, food, transportation, etc.). The interconnectedness of the flows that link us is more important than many realize, as they sit in their home, car, workplace, or restaurant—its potential economic savings and carbon-mitigation benefit far surpass that of incremental efficiency gains. The NZCC approach that follows leverages this connectedness (as well as old-fashioned efficiency) for cost-effectiveness and repeatability. Not only are the steps sequenced in terms of their cost-effectiveness, each step generally improves the performance of a later step (e.g., not only is efficiency cheaper than new renewable supplies, it also means less renewable energy supplies are needed; not only are properly sited and sized buildings energy-efficient in themselves, they also mean that efficient transport systems are more likely to be cost-effective). The result of this synergy between steps is greater whole cost effectiveness than cost effectiveness of any individual step taken by itself.

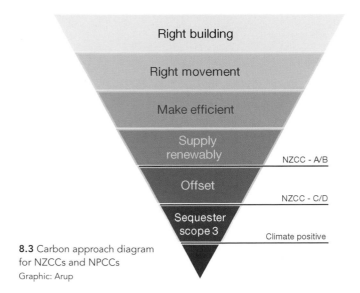

8.3 Carbon approach diagram for NZCCs and NPCCs
Graphic: Arup

Of course, the idea of a shed is just a subcomponent in a larger cycle: the carbon cycle. So what happens when the flow reaches its final step? It returns to the first step and recycles until the whole achieves the desired balance.

A Comprehensive and Prioritized Approach

Following the approach outlined in Chapter 6, the comprehensive, prioritized, and repeatable NZCC approach is detailed below, followed by an in-depth discussion of optimizing systems at the community scale.

1.   Right Building

Size: Since 1950, the size of new homes per capita has increased up to 300 percent in countries around the world, with the largest average home size of 243 square meters (2,616 square feet).[14] These larger buildings mean

– More wall area to lose (or gain) heat

– More space to artificially light
– More volume to ventilate
– More materials to construct (with embodied carbon in their manufacture)
– More goods to accumulate (with embodied carbon in their manufacture).

Location: By locating new developments in well-planned communities and infill sites, infrastructure and transit system burdens are lessened.

Shared walls: By encouraging multistory and dense development, a greater percentage of occupied space will share walls or indirectly buffer adjacent buildings due to their proximity.

Flexibility, resilience, and embodied carbon: Ironically, few new buildings are built with future users in mind. Others lack the resilience needed to be reoccupiable after an earthquake, hurricane, or flood. The lack of flexibility and resilience results in short-life structures that are often landfilled within twenty to thirty years of their original ribbon cutting. Since the embodied carbon in a building can be a significant part of its life-cycle embodied carbon, this oversight is both an opportunity for carbon mitigation *and* an opportunity for better financial protection of assets.

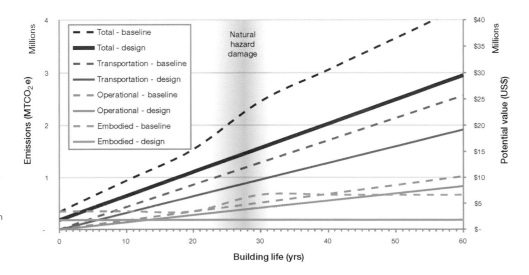

8.4 Lifetime carbon emissions of a sample building—transportation is often the largest cause; operational and embodied carbon typically cross at approximately ten to twenty years and can be significantly affected by resilience and efficiency
Graphic: Arup

2. Right Movement

Movement is often the largest or second largest source of emissions within a community. Movement-related emissions come predominantly from all of the moving about to satisfy our *needs*, not our *wants*. Most people don't want to drive to work; they need to get there. They don't want to drive to the store; they need to make a purchase. They don't want to drive to dinner; they need to eat out. So what better way to reduce movement emissions than to bring needs closer together while making access to those needs more effective and pleasant to experience?

Pedestrian-oriented: By creating complete communities that provide freedom of choice for citizens to move about, people can choose walking and biking. Face-to-face relationships are developed and community health improves (reducing health-care costs significantly).

Mixed-use: By providing access to community-supporting services such as restaurants, local retail, banking, child care, laundry, libraries, pharmacies, postal services, and schooling, vehicle trips are reduced, community is created, and public health improves.

Transit-friendly with car sharing: By reducing the amount of land given over to cars and increasing the amount of land given over to housing and open recreational space, quality of life increases while providing the needed density for transit effectiveness. For trips that aren't suitable via transit, car sharing is available and offers not only reduced carbon emissions but the added benefits of included maintenance, fuel, and insurance.[15] Those who still choose to own private automobiles (e.g., for transporting children or distance commuting) end up having less congested roadways and *better* access to parking since fewer cars are on the street.

Digital presence: Some of the best modern methods of reducing movement are the digital services of mobile computing, video calling, and cloud computing—each of which offers faster and more economical means of communication. Of course, person-to-person contact in buildings is still needed, which brings us to Step 3.

3. Efficient Movement and Buildings

Movement and buildings are necessary and important facets of our economy and society. Just consider how much more effective it is to meet in person first and then follow up with a phone call, than it is to start with a "cold" phone call. So when buildings and movement are needed, Step 3 encourages that they be made efficient and cost effective.

Building efficiency

The NZCB six-step approach discussed in Chapters 6 and 7 details how to make buildings efficient through (1) load reduction, (2) passive systems, (3) active efficient systems, (4) energy recovery, (5) building integrated generation, and (6) offsets. In this chapter, we'll focus on transportation efficiency and community-scale system efficiency.

Transportation efficiency

Mass transit: Transit system efficiency can be considered in two ways: the efficiency per unit of distance traveled (i.e., vehicle or system efficiency) and the efficiency of the system per *passenger unit* of distance traveled. Both are good metrics for improvement, with the latter often resulting in a larger *net* decrease in carbon emissions. The irony is that a focus on transportation efficiency per passenger distance traveled often results in a gross carbon *increase* for the transit system (as ridership increases) while reducing net carbon emissions in the community (as single occupant vehicle trips are eliminated). A cobenefit of mass transit is the cost saving to commuters. Daily commuters typically save over U.S.$10,000 per year in major cities when compared to driving an automobile.[16] The approach steps for improving transportation system carbon performance is to (1) increase ridership, (2) decrease weight/rolling resistance, (3) install efficient systems, (4) recover energy, (5) renewably source and (6) offset.

Automobiles: Although vehicle efficiency largely stagnated after the 1970s oil crisis, recent developments in electric vehicle technology and policy are resulting in significant improvements in vehicle efficiency per mile traveled.[17] When combined with car-pooling activities, the efficiency per passenger mile traveled is increasing even faster. Hence the smartest cars may not be the self-proclaimed

Energy Recovery at Campus Scale: Stanford University

In 2009, Stanford University analyzed its heating and cooling systems, not separately, but together. The analysis was clear. When some operations on campus needed heat, other processes were rejecting it. The resulting "simultaneous load" insight has led the university to invest millions over five years to install more than ten miles of new piping, retrofit all building connections, and construct a new central energy facility with energy recovery chillers. The resulting savings are projected to exceed $600 million over 40 years, 50 percent of Stanford's annual GHG emissions, and 18 percent of campus water use.[18]

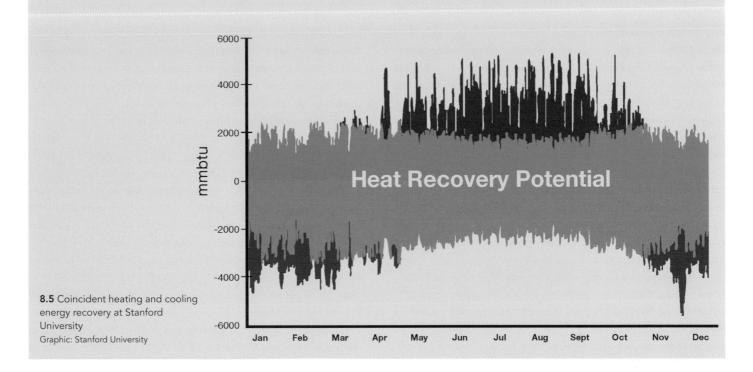

8.5 Coincident heating and cooling energy recovery at Stanford University

Graphic: Stanford University

two-seat Smart cars, but rather the four-seat diesel Volkswagen Jettas (over 180 miles per gallon per passenger potential) and the plug-in Toyota Priuses (over 200 miles per gallon per passenger potential). Equally exciting are the developments in pure electric vehicles such as the Tesla Model S sedan, which appears better in nearly every way (when compared to other $50,000+ luxury cars).[19] Add to these impressive vehicle developments the increasing potential to realize autonomous driving in the next two decades. Such computer-driven vehicles promise not only greater safety through faster response times, but also the potential for "car-train lanes" that would displace our carpool lanes in energy efficiency and priority. Such lanes would allow more compressed vehicle spacing (less roadway congestion), improved aerodynamic drafting potential, improved safety, and hands-free driving so occupants can talk on their phones and text to their hearts' delight.

Humanizing the street: The benefits of improved movement efficiency can significantly improve the human friendliness of the street. Local carcinogens and particulates will decrease, and noise will be reduced. This has clear benefits to human health. Additionally, the quiet and pollution-free (at point of use) vehicles will allow operable windows to be used and air-conditioning systems to be turned off.

Community-scale system efficiency

A great deal of attention has been paid to "green" buildings—first in the 1970s, then again in the first decade of the twenty-first century. Although laudable, this emphasis has sometimes missed the greater opportunity of community-scale system efficiency. Individual buildings have installed their own wastewater treatment, composting, cogeneration, thermal energy storage, chilled water, hot water, and renewable energy systems. Often these systems are appropriate and effective. On other occasions they are a missed opportunity for more cost-effective and efficient system opportunities at scales beyond an individual building (e.g., due to load sharing among mixed-use developments). The latter portion of this chapter addresses the optimization of these community-scale systems.

4. Renewable Energy Supply

Once (1) the right buildings are built, (2) the right movement is occurring, and (3) both are efficient, then renewable fuel sources are added. For buildings, this takes the form of distributed and utility-scale renewable electric and thermal energy systems. For movement, this takes the form of biofuel sources and decarbonized electrical supplies (from the distributed and utility-scale electric energy systems that can also serve buildings).

The community scale offers advantages as economy of scale and minimum system sizes emerge, which are otherwise unavailable at the small scale of individual buildings and vehicle owners. Such emergent possibilities include waste to energy/biofuel, biomass gasification, wind turbines, solar thermal electric systems, and concentrating photovoltaic systems, to name just a few.

Once renewable energy is supplied, an NZCC achievement is possible, but only if the renewable energy supply is enough to accommodate all community carbon emissions. If not, more attention is needed.

The optimization of community-scale renewable generation (or alternatively "decarbonized generation") is discussed in further detail in the latter part of this chapter.

5. Offsets

If community renewable energy sources are inadequate to meet all needs, offsetting can be used to achieve an NZCC designation. Offsetting has already been covered in Chapter 3 and can include direct access power agreements and tradable certificate purchases. When applied to community-scale efforts at carbon emissions reduction, the conversation typically transitions in a few minor but tangible ways. First, the amount of offsets is increased since the additional carbon sources (e.g., transportation) are included. Second, the cost of offsets is sizably *reduced* since the lower administrative costs of buying and selling result in a beneficial economy of scale.[20] Third, there is often a heightened desire to localize offsets for greater emotional resonance. As a result, a community may reach out beyond the boundaries of its carbon accounting for local offset projects.

6. Sequestration and Scope 3

Once (1) the right buildings are built, (2) the right movement is occurring, (3) both are as efficient as possible, and renewable fuel is supplied (4) on-site or (5) off-site, communities have the broader power to sequester carbon or shift consumer choices, thereby achieving an NPCC designation.

Sequestration

Much of the world's carbon is bound in the world's biomass. This biomass includes not only trees, shrubs, and *you*, but the very soil beneath, which (when healthy) is alive with biological activity. Sequestration in this biomass may take the form of a landscape action, a restoration action, or a preservation action.[21]

Landscape and the urban forest: New and existing communities have an opportunity to plant trees, add vegetated rooftops, and open up sidewalks to bioswales. In Manhattan, a dilapidated rail line has been restored as an urban park. In Sacramento, California, the city aims to double its claim to the most trees per capita and plant five million more trees by 2025.

Restoration: As communities age, they have opportunities to reverse the damage of their younger, wilder days. This may take the form of artificial reef planting, restored estuaries, daylighted streambeds, and managed flooding of riverbeds. In the case of communities that are decreasing in population such as Detroit, Michigan, and New Orleans, Louisiana, there is the potential for transformational change as large tracts of built urban form are replaced with greenbelts and parkland. All such restoration builds biomass while improving community health.

Preservation: One of the greatest benefits of right building and movement is the building and infrastructure development that *does not* occur on farmland, grassland, wetland, or forested land. Since communities typically exist in lowland areas that are relatively flat, their edge growth tends to consume the most biologically rich and agriculturally valuable land. In just the last ten years, California, the leading producer of agricultural products by value in the United States, has lost 6 percent of its prime farmland to development. If such rates continue, no prime farmland will remain by the middle of the next century.[22] As a result, a cobenefit of carbon mitigation is the assurance that we'll have adequate farms to feed us in the future. No farms, no food.

8.6 The optimal scale of energy, water, and waste systems can vary due to many factors, not least of which is scale
Graphic: Arup/Cole Roberts

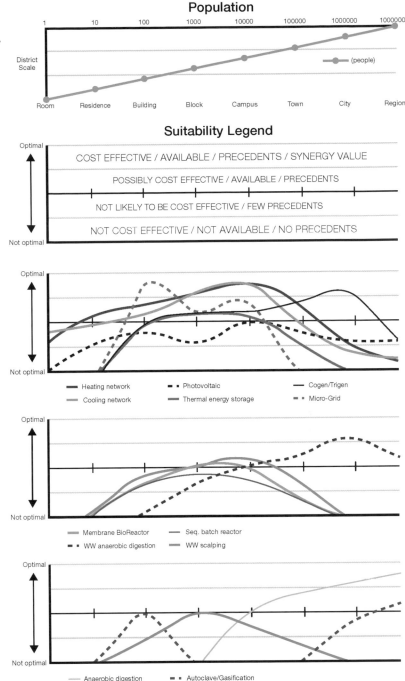

Consumer consumption shifts

As discussed in the "Community Consumption of Goods and Services" box earlier in this chapter, the consumption of goods and services within a community and other Scope 3 sources of emissions can be a large opportunity for climate-positive actions. Examples of such actions can include programs that support emissions reductions by other organizations (Scope 1 and 2 emissions of others) and the development of sustainable food systems, product labeling, and zero-waste packaging policies.

DECARBONIZING COMMUNITY SYSTEMS

The Optimal Scale of Systems

If we study the systems in our communities, we see the scale at which they have *historically* found their optimum. Their resting places have been determined by complex factors similar to those that shape our infrastructure today: politics, economics, finance, efficiency, expertise, law, and custom. They are not necessarily at the optimum scale for climate-change response or cost effectiveness, but understanding them is helpful. By studying their profile, we can start to ask how the reflected systems can more often be implemented at the appropriate scale. In Figure 8.6, we have taken a selection of systems broken down by energy, water, and waste. Not all systems are shown, and the profiles represented are not true for all occurrences. However, they are a reasonable approximation and show, for example, the importance of a population greater than 100,000 people for wastewater-anaerobic-digestion plants to pencil out, the increasing economy of scale that supports district heating and cooling systems, and the microgrid that is most suitably implemented at scales between 100 and 100,000 people.

Creating Low-Carbon Community Systems

Low-carbon community systems don't just happen overnight. They take time, process, and new tools.

1. Establish, Expand, Optimize, Maximize

The elements of time and phased growth are very important in community-scale system development. Typically, a system is established (or piloted) at the minimum scale necessary for cost effectiveness. As demand for the system and funding becomes available, it is then expanded to provide service to the new users. As it expands and grows in size, it is optimized for better and better performance. Lastly, it finds its maximum potential and exists in service for as long as the community exists or finds it beneficial.

This growth framework means that sometimes there is no system at the beginning, but rather a collection of smaller systems designed to be able to connect as a community-level system in the future. Such a nodal growth was planned for Nevada State College in 2005. Recognizing the college's limited funding, a central plant was not viable in the first phase. Instead each building was planned for future interconnection in a networked system.

Perhaps the most classic example of the "establish, expand, optimize, maximize" principle is the growth of a transit system. Initial routes are almost always heavily subsidized, under-ridden, and small in scale. As they expand and optimize over decades, they often become integral and critical parts of the community.

It's important to recognize that optimization not only applies to the efficiency of the system itself, but is interdependent with other elements in the district. This interrelationship and synergy is the cornerstone of an ecological district (or *ecodistrict*), as one element in the district has a positive impact on another.

The bus rapid transit system in the Brazilian city of Curitiba is well known around the world as a model of cost-effective mass transit. Nearly 1.3 million commuters ride a system that cost 90 percent less to build than the next best alternative. It is such a model that it has spurred an eagerness in other communities to implement similar systems. Although there are successes, to this day, no other community has come close to the performance of the Curitiba system. The reason for its success is the synergistic elements of the transit ecodistrict, which is much more than just good buses in dedicated lanes.[23] This took time to create.

- When the bus rapid transit system was first ESTABLISHED in Curitiba in 1978, it was just 2 percent of its current size.
- The early success of the bus rapid transit justified its EXPANSION. As new routes were added, new buses replaced the old ones and additional stops were added. And the new buses and stops were not necessarily identical to their predecessors.
- As they expanded, the city buses and stops were OPTIMIZED. Loading and unloading patterns, stop frequency, average speed, and length of route resulted in changing shapes, engines, and boarding areas. The wayfinding system was improved so that visitors and new users could navigate the system with confidence. Even greater benefit resulted from concentrated development along the transit corridors. The building occupants benefited from access to fast and cheap transit. The transit system benefited from ever-increasing ridership.
- Today the bus rapid transit ecodistrict is arguably MAXIMIZED. It is the lifeblood of the city and a model around the world.

2. New Optimization Tools

Integrated Resource Management (IRM, aka integrated carbon management) was first introduced in concept in Chapter 4. In action, IRM and other tools like it have shown their value in extracting the significant synergy that can exist between energy, water, waste, transit, etc. In a recent community IRM application, the following synergies were just a few of those that arose from the model; each was *in addition to* the benefit of the original efficiency strategy and therefore would have been unaccounted in prior community infrastructure planning:

8.7 Curitiba bus rapid transit ecodistrict
Image: Curitiba Municipal Archives, adapted by Arup

- Water efficiency of 15 percent yielded 3 percent energy savings.
- Energy efficiency of 25 percent yielded 40 percent water savings, which yielded an indirect 4 percent additional energy saving.
- Electric vehicle strategies yielded 3 percent carbon reduction, 10 percent parking reduction, and 6 percent increased energy demand.
- Waste diversion to anaerobic digestion yielded a 5 percent renewable energy production source.

Equally exciting is the ability of tools like IRM to forecast performance based on alternative strategies and link the forecasts with the "establish, expand, optimize, maximize" rollout over the community's phased development. The same project that exhibited the synergies above also analyzed alternative good, better, and best strategies as applied to the community's planned growth. The output quantified the impacts of the strategies and established a planned performance curve that could be used to check absolute progress as development occurred (Figure 8.8).

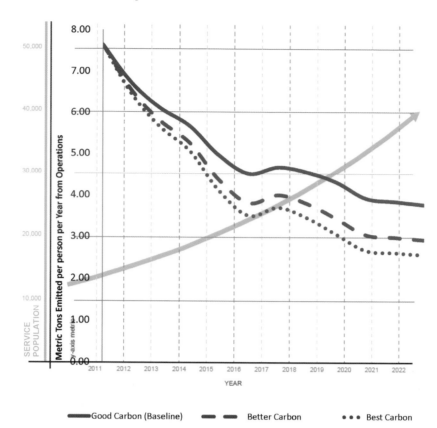

Density Scenario 1

━━━ Good Carbon (Baseline)   ▬ ▬ Better Carbon   ● ● ● Best Carbon

Density Scenario 2

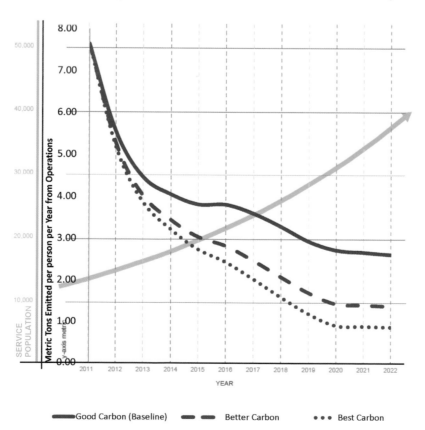

━━━ Good Carbon (Baseline)   ▬ ▬ Better Carbon   ● ● ● Best Carbon

3.   New Financing and Procurement Options
One of the greatest drivers of community-scale infrastructure costs, and often the greatest obstacle to implementation, centers on one word: risk. This may be perceived risk that is not real. It can also be real risk that is poorly managed.

Historically, the procurement process for new infrastructure concentrated the risks with one party (typically a public agency or private institution), or with parties that were not willing or able to manage the risk. Today, new options are emerging that offer the opportunity for risks to be borne by those parties best able or willing to manage them. The result is a variety of procurement routes that weren't available decades ago. There are two significant benefits offered by these routes:

1. Improved risk management can mean lower total cost of development.
2. Third-party financing can mean that debt is converted to operating expense, which can improve the balance sheet.

Public–private partnerships (aka PPP or P3) are showing up not only in the historic arena of highway development, but in community energy, water, and waste system development. Managed energy service agreements and independent power providers are now contracted in multiple communities around the world, and many of these are self-financed. These new procurement options are not a panacea for old impasses, but rather new tools in the toolbox—tools that can improve the likelihood of progressing community-scale systems beyond the stage of just a good idea and into the realm of today's ecological infrastructure.

LOW-CARBON AND CLIMATE-POSITIVE COMMUNITIES IN SUMMARY

The interdependency and economy of scale that occurs beyond the scale of individual buildings is often where some of the greatest opportunities exist.

NZCCs are complicated by the need to consider additional emissions sources, diverse ownership interests, and limited access to data; the presence of both new and existing buildings; and the extended time horizon of community carbon-reduction strategies.

The definition of an NZCC is "[a community] that has greatly reduced energy needs through efficiency gains such that the balance of energy for vehicles, thermal, and electrical energy within the community is met by [on-site or off-site] renewable energy." The definition goes on to further "grade" NZCCs from A to D based on their renewable energy supply.

An NPCC is a community that achieves NZCC and goes further to reverse historical carbon debt by sequestering carbon at an annually positive rate for the life of the community and/or measures and reduces Scope 3 emissions using established accounting protocols.

Once an understanding of a community's emissions and NZCC goal are set, there is still a need for a cost-effective and repeatable approach to get to zero. The steps detailed include the following:

1. Right building: size, location, shared walls, flexibility, resilience, and embodied carbon
2. Right movement: pedestrian-oriented, mixed-use, transit-friendly, with car sharing and digital presence
3. Efficient movement and buildings: building efficiency (Chapter 7), mass transit and automobile efficiency, humanizing the street, and community-scale system efficiency
4. Renewable energy supply: for buildings, renewable electric and thermal energy systems; for movement, biofuel sources and decarbonized electrical supplies—once renewable energy is supplied, NZCC achievement is possible
5. Offsets: if community renewable energy sources are inadequate to meet all needs, offsetting can be used to achieve net zero
6. Sequestration and Scope 3: communities have the broader power to sequester carbon or shift consumer choices, thereby becoming an NPCC

Low-carbon community systems take time, process, and new tools. They must be established, expanded, optimized, and maximized over decades.

New optimization tools are emerging, such as IRM. The value of such analysis includes greater understanding of the synergy between resources, phasing impacts of strategies, and the ability to explore questions through computational iteration. In addition to new modeling tools, new financing and procurement tools are emerging to better manage risk and identify sources of financing.

NOTES

1 New York: Random House, 1961.

2 Two percent of global GDP is a revised estimate of the Stern Review (*Stern Review: The Economics of Climate Change*, 2006, http://webarchive. nationalarchives.gov.uk/+/http:/www.hm-treasury.gov.uk/independent_ reviews/stern_review_economics_climate_change/stern_review_report.cfm [accessed June 8, 2012]). Two percent of 2010 global GDP is approximately U.S. $1.2 trillion (World Bank, "World Development Indicators," http:// data.worldbank.org/data-catalog/world-development-indicators [accessed June 2012]).

3 For the purpose of clarity, *NZCBs* and *NZCCs* used in this text are synonymous with the U.S. National Renewable Energy Laboratory definitions for Net Zero Energy[Emissions Basis] buildings (NZEBs) and Net Zero Energy[Emissions Basis] communities (NZECs).

4 Fugitive emissions result from intentional or unintentional releases of greenhouse gases, including the leakage of hydrofluorocarbons from refrigeration and air-conditioning equipment and the release of methane from institution-owned farm animals.

5 P. Erickson, M. Lazarus, E. A. Stanton, and F. Ackerman, *Consumption-Based Greenhouse Gas Emissions Inventory for Oregon*, Stockholm Environmental Institute report commissioned by the Oregon Department of Environmental Quality, 2011.

6 Christopher M. Jones, Daniel M. Kammen, and Daniel T. McGrath, *Consumer-Oriented Life Cycle Assessment of Foods, Goods and Services*, Energy and Climate Change, The Berkeley Institute of the Environment, UC Berkeley, 2008, http://escholarship.org/uc/item/55b3r1qj (accessed June 8, 2012).

7 Scope 1 refers to direct greenhouse gas emissions occurring from sources that are owned or controlled by the community, including in-community stationary combustion of fossil fuels, mobile combustion of fossil fuels by community-owned or community-controlled vehicles, and fugitive emissions. Scope 2 refers to indirect emissions generated in the production of electricity consumed by the community. Scope 3 refers to all other indirect emissions—those that are a consequence of the activities of the community but occur from sources not owned or controlled by the community. Typical inventories account for only Scope 1 and 2 sources; however, a new standard and guidelines for Scope 3 emissions reporting has been drafted by the World Resources Institute and World Business Council on Sustainable Development, the original authors of the Greenhouse Gas Protocol (http://www.ghgprotocol.org/). U.S. Environmental Protection Agency greenhouse gas accounting procedures require reporting only Scope 1 and 2.

8 Chuluun Togtokh, "Time to Stop Celebrating the Polluters," *Nature* 479(7373), November 16, 2011.

9 Nancy Carlisle, Otto Van Geet, and Shanti Pless, *Definition of a "Zero Net Energy" Community*, NREL/TP-7A2-46065, Golden, CO: National Renewable Energy Laboratory, November 2009.

10 Shanti Pless and Paul Torcellini, *Net-Zero Energy Buildings: A Classification System Based on Renewable Energy Supply Options*, NREL/TP-550-44586, Golden, CO: National Renewable Energy Laboratory, June 2010.

11 Nancy Carlisle, Otto Van Geet, and Shanti Pless, "Definition of a Zero Net Energy Community," National Renewable Energy Laboratory, Technical Report, NREL/TP-7A2-46065, 2009.

12 An example is Climate Positive (http://climatepositive.org/) in Australia, which in addition to advocating efficiency and carbon offsetting also provides a path for individuals and businesses to take a climate-positive step through biomass restoration projects.

13 An example is One Planet Living (http://www.oneplanetliving.org/), a global initiative based on ten principles of sustainability developed by BioRegional and the World Wildlife Fund. For example, One Planet communities such as BedZED are challenged to address Principle 5: "Choosing low-impact, local, seasonal and organic diets and reducing food waste."

14 R. Diamond and M. Moezzi,"Changing Trends: A Brief History of the US Household Consumption of Energy, Water, Food, Beverages and Tobacco," Lawrence, Berkeley National Laboratory, n.d. http://epb.lbl. gov/homepages/rick_diamond/LBNL55011-trends.pdf (accessed June 8, 2012); Simon Johanson, "Australian Homes Still the World's Biggest," *The Sydney Morning Herald*, August 2011.

15 Car-sharing programs such as Zipcar reduce carbon emissions by charging per use. The psychological impact of such "pay per use" or service programs is a net dissuasion compared to the "sunk cost" ("use it since you bought it") psychology of vehicle ownership. There is some rebound

effect since the amenity value of a vehicle, even a pay-per-use vehicle, is great enough to overcome the cost and therefore allow amenity use (drive to the countryside) that would not otherwise occur if no vehicle option existed.

16 American Public Transport Association, "Commuters Who Resolve to Save Money in 2012 Take Note: Transit Riders Save More as Gas Prices Increase," press release, 2012, http://www.apta.com/mediacenter/pressreleases/2012/Pages/120111_TransitSavingsReport.aspx (accessed February 1, 2012).

17 California Environmental Protection Agency Air Resources Board, "Clean Car Standards—Pavley, Assembly Bill 1493," 2010, http://www.arb.ca.gov/cc/ccms/ccms.htm (accessed June 8, 2012).

18 For more information, visit Stanford's sustainability portal at http://ssu.stanford.edu/heat_recovery.

19 See www.teslamotors.com.

20 Offsets for a single resident purchaser are typically over 1 cent per kilowatt hour. Offsets for a large portfolio or municipal purchaser are typically less than 0.1 cent per kilowatt hour.

21 Note that we consider the majority of carbon capture and sequestration activity to be a decarbonization of the electrical supply, and hence included in Step 3, 4, or 5, depending on the accounting boundary.

22 California Department of Conservation, Farmland Mapping and Monitoring Program, 2008. Electronic files, GIS data, and fact sheets available at http://www.conservation.ca.gov/dlrp/fmmp/Pages/Index.aspx.

23 For more information on Curitiba's bus rapid transit ecodistrict, visit http://www.fta.dot.gov/4391.html.

The Old Mint, originally constructed in 1874, is planned for renovation as a
twenty-first-century Zero Net Energy LEED Platinum "Gateway to San Francisco"
Image: HOK/Arup

9

Getting to Zero for Existing Building Stock

Fiona Cousins

The Darwinian mechanism of vary-and-select, vary-and-select has one enormous difference from the process of design. It operates by hindsight rather than foresight. Evolution is always away from known problems rather than toward imagined goals. It doesn't seek to maximize theoretical fitness; it minimizes experienced unfitness. Hindsight is better than foresight. That's why evolutionary forms such as vernacular building types always work better than visionary designs such as geodesic domes. They grow from experience rather than from somebody's forehead.

Stewart Brand, *How Buildings Learn*[1]

So far, we have focused on new buildings. However, in Europe and North America, especially in existing cities, the carbon emissions associated with existing buildings are much greater than those associated with new buildings. This is mostly because the amount of area contained in existing buildings is much larger than the area in new buildings. Estimates for the United States show that only 27 percent of the existing building stock will be replaced between 2005 and 2030, which means that, of the buildings that we will be using in 2030, most already exist.[2] In China, India, and Brazil, a smaller proportion of the future building stock already exists because of the rapid rate at which the building fabric is increasing, but existing building energy use is still an important factor in overall emissions.

If we are to reduce the carbon emissions associated with buildings to the levels needed to maintain environmental carbon dioxide levels at 450 parts per million, we have to reduce the emissions associated with existing buildings.

The opportunity and need for reducing energy use and carbon emissions in existing buildings are variable. The way that energy is used in buildings is described in detail in Chapter 7. In commercial buildings, the predominant uses are for lighting, cooling, and heating, and in residential buildings, the major uses are for heating, cooling, and hot water heating.

Very old commercial buildings sometimes have very good performance because they were designed to operate without air-conditioning and to maximize the use of daylight. These buildings typically have narrow plans, recessed windows, high ceilings, and limited glazing. However, there are also buildings that were built

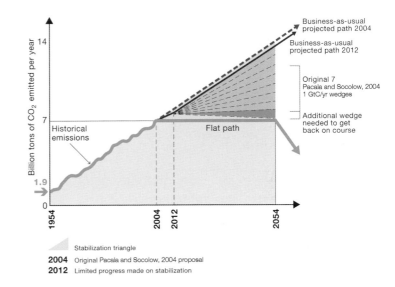

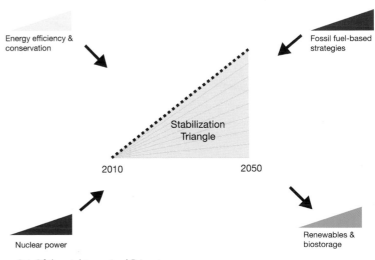

9.1 Of the eight required Princeton Stabilization Wedges, one is attributed to carbon emissions reductions associated with energy savings of 25% in buildings
Graphic: Arup, Data: Princeton University Carbon Mitigation Initiative

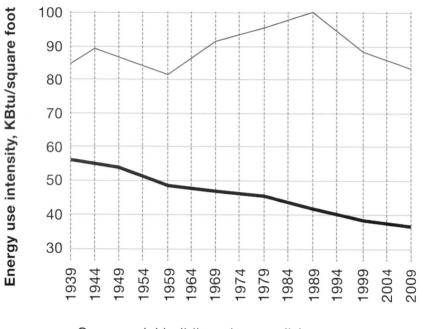

9.2 Average energy use intensity for U.S. commercial and residential buildings over time; note that as homes have become larger, EUI has decreased
Graphic: Arup, Data: U.S. DOE CBECS Survey 2003 and McKinsey & Company

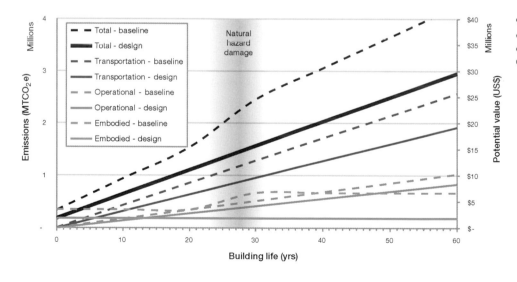

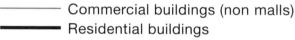

9.3 The relative size of embodied carbon emissions and operational carbon emissions over time
Graphic: Arup

when energy was cheap and glazed facades were fashionable. These buildings typically have high energy use.

It is inconceivable that the entire existing building stock will be replaced in the near future. In addition to the economic constraints, many buildings continue to fulfill useful social and economic purposes and some buildings are also part of our cultural heritage. Replacement also carries a heavy cost in terms of embodied carbon emissions.

The effect on carbon emissions of retaining existing buildings can be considered in two ways: either the carbon emissions associated

with construction of existing buildings can be considered as a "sunk" cost or the total carbon emissions associated with the construction can be "amortized" over the total life of the building, giving a lower per-year number if the building is retained. In either case, the effect of retaining existing buildings has a carbon benefit.

This chapter focuses on the potential for carbon emissions reductions within existing buildings and describes some of the barriers to and motivations for making these changes.

9.4 Re-skinning projects such as this one provide the opportunity for replacement of both systems and envelope; renovations on this scale can be expensive and complex
Images: Arup/Daniel Imade

One of the major considerations when implementing carbon emissions reductions in an existing building is that it has occupants. There are two parts to this: firstly, the occupants have an enormous influence on the amount of energy used within a building, and occupant education and engagement can be a key component of a carbon emissions reduction program.[3] Secondly, the occupants will restrict the ease of access for maintenance or construction of systems. In office buildings, occupant salaries are the major cost for the business, at approximately 50 to 100 times the cost of the energy used, so energy cost is rarely a "top of mind" issue for occupant

WHAT'S DIFFERENT ABOUT EXISTING BUILDINGS?

For a new building, the design team can make decisions about every aspect: the orientation, massing, envelope, and systems. Once the building has been constructed, it is much more expensive, and therefore difficult or impossible, to change these decisions. The focus for reducing carbon emissions in existing buildings is therefore to make the best of the existing conditions and systems.

This means that existing-building strategies tend to focus on the operation of the building and its systems. At points of major renovation—for example, when the building envelope or major equipment reaches the end of its useful life—it may be possible to make significant system changes. However, there is strong interdependence between the components and they have different life expectancies, so wholesale changes usually occur only when a building undergoes a substantial renovation, such as at ownership or occupancy changes.

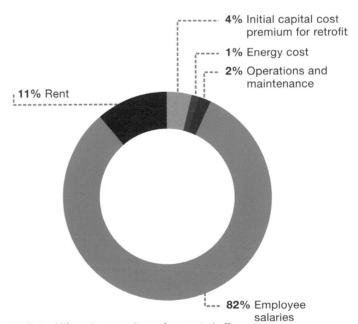

4% Initial capital cost premium for retrofit

1% Energy cost

2% Operations and maintenance

11% Rent

82% Employee salaries

9.5 Typical life-cycle expenditures for a typical office
Graphic: Arup, Data: RetroFit Depot[i]

9.6 Carbon emissions abatement curve. Green bars indicate measures that may be taken as part of work in commercial or residential buildings
Graphic: Arup, Data: McKinsey & Company[ii]

Cost   Real 2005 dollars per ton CO$_2$e

Potential   Gigatons/year

Commercial electronics
Residential electronics
Residential buildings - lighting
Commercial buildings - LED lighting
Fuel economy packages - cars
Commercial buildings - CFL lighting
Residential buildings - new shell improvements
Fuel economy packages - light trucks
Commercial buildings - new shell improvements
Commercial buildings - combined heat and power
Cellulosic biofuels
Industrial process improvements
Industry - combined heat and power
Existing power plant conversion efficiency improvements
Residential water heaters
Conservation tillage
Coal mining - methane management
Commercial buildings - control systems
Manufacturing - HFCs management
Residential buildings - shell retrofits
Nuclear new-build
Onshore wind - low penetration
Natural gas and petroleum systems management
Active forest management
Afforestation of pastureland
Reforestation
Winter cover crops
Onshore wind - medium penetration
Distributed solar PV
Coal power plants - CCS new-builds with EOR
Biomass power - cofiring
Coal power plants - CCS rebuilds with EOR
Onshore wind - high penetration
Afforestation of cropland
Commercial buildings - HVAC equipment efficiency
Coal power plants - CCS new-builds
Solar CSP
Industry - CCS new-builds on carbon-intensive processes
Residential buildings - HVAC equipment efficiency
Coal power plants - CCS rebuilds
Coal-to-gas shift - dispatch of existing plants
Car hybridization

firms. Occupants are also the only source of complaints about the building systems, and their hot/cold complaints can lead to systematic energy use increases.[4] In residential buildings, occupant behavior is the most important factor in carbon emission reductions.

WHAT'S THE OPPORTUNITY?

There has been a lot of research into the big-picture strategies that will allow us to achieve a carbon emissions rate that maintains an environmental carbon dioxide level of 450 parts per million or less, as shown in Figure 9.1. This research stresses the importance of existing buildings, both commercial and residential, to achieving these targets. McKinsey & Company, a global management consulting firm, carried out an economic analysis, shown as an abatement curve in Figure 9.6, which outlines the most cost-effective strategies for achieving carbon emissions reductions. In this diagram, the width of each bar indicates the cost and the deepest bars have the most significant impact. It is clear that energy-efficiency retrofit, especially in residential projects, can have a big impact.

While the abatement curve indicates the places where we should look for inexpensive carbon abatement measures, it is very general. Our project experience shows that the opportunities for carbon emissions reductions and the cost and carbon payback for each potential measure are very dependent on the following circumstances:

- The climate—a measure that pays back rapidly in a cold climate might never achieve break-even in a warm one
- The way in which the building is operated, including whether there is a facilities manager, how well-educated she is in energy matters, and what her incentives are
- The relationship between the facilities manager and the occupants
- The way in which the operations and maintenance expenditure for a facility is managed
- The type of systems installed in the building
- Local utility costs
- How well the building works for its current purpose
- The expectations of the occupants.

It is important to note that reducing carbon emissions can also benefit occupants and building owners immediately: the best way for these groups to reduce carbon emissions is to save energy, which almost always saves money.

TECHNICAL APPROACH
Benchmarking

In many areas of life, the first step in resolving an issue is to accept that you have a problem. Similarly, the first step in reducing energy use in buildings is to understand how much energy is being used and what the major energy uses are. This data can be used to identify how big a problem a particular building has.

In existing buildings, there is usually real energy data from energy bills. Unfortunately, often the only data available is the total for the building, which is not broken down by end use. But, even this limited data is useful because it can be compared with national and regional energy benchmarks to show whether the building is a good or bad performer compared to similar building stock.

In the United States, these comparisons can be achieved for commercial buildings through the use of the Environmental Protection Agency's Portfolio Manager tool or through the use of the Department of Energy's Commercial Buildings Energy Consumption Survey (CBECS) database. In Europe, energy in use for public buildings is generally published through a Display Energy Certificate, as required by the European Energy Performance of Buildings Directive (EPBD), which gives a grade for each building (see Chapter 3). In Europe, residential energy certificates are also available and are required when renting or buying a property. In Australia commercial buildings have been required to disclose their energy efficiency to potential buyers and tenants since 2010.[5] Benchmarking is such a powerful tool that some cities in the United States are starting to require it.[6]

If more detailed information is available, it can help evaluate the areas of greatest energy use. At the simplest level, this might mean that a building is metered floor-by-floor or tenant-by-tenant. If a building is metered in such a way and the activities on each level are generally similar but energy use is different, this can help to pinpoint areas of waste. It can also be used to set up competition between floors or tenants. A technique like this has been used at a number of university campuses where different halls of residence and different floors within them have been asked to compete to reduce their energy use.[7] We have even run a competition between the floors of the Arup San Francisco office. Some utilities provide residential customers with information about how their bills compare to their neighbors', which can be helpful in allowing individual households to recognize that they could do better.[8]

Not all buildings have floor-by-floor or tenant-by-tenant metering. In this case, utilities are usually factored into the rent, based on area or some other arrangement. Most of these arrangements mean that low-intensity users subsidize high-intensity users and also remove any incentive for individual tenants to reduce their energy use. As a result, some states are planning to introduce legislation to make direct individual metering compulsory for tenants.[9]

Measuring, Monitoring, and Managing

It is a management truism that only what is measured is managed, and it is true of energy use as well. The initial measurements are used to set up a baseline for the effects of monitoring and managing

energy use. The simplest baseline is to use whole-building energy data, for example energy bills, and to compare energy use across time—for example, comparing bills for January over successive years. If the bills differ and the weather was not unusual, this can indicate that there are problems with the systems themselves or with the way they are being operated. In small buildings, it will probably be easy to identify where the problem is; in large buildings, additional monitoring and analysis is probably required.

The simple act of monitoring the rate of energy use (power) is enough to lead to significant, low-cost carbon emissions reductions: installation of real-time power monitoring in 30,000 homes in Ontario showed that these led to energy-use reductions of up to 15 percent with an average energy-use reduction of 6.5 percent.[10]

In complex buildings monitoring is usually provided through a building management system (BMS). A BMS usually controls all the HVAC equipment, managing processes like turning equipment on and off, and controlling valves, pumps and fans to deliver the desired room temperature. To do this it monitors equipment operating status and a large number of sensors, and then uses its internal

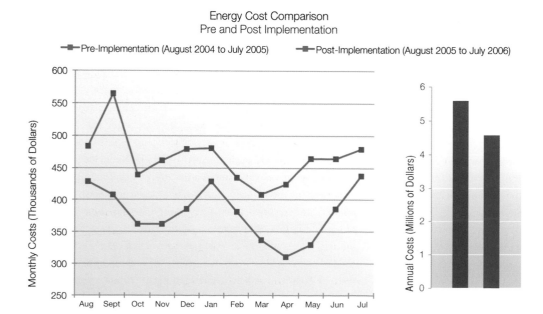

9.7 Energy use comparison over two years of a project; the second year followed significant energy saving initiatives
Graphic: Arup

programming to change the system operation. A BMS can also be used to record the status of equipment or sensors at specified time intervals, a process called *trending*. The trends can be used to identify out-of-control equipment, changes over time, or errors, and are an essential tool in retrocommissioning. The appropriate baseline can be either the behavior predicted by the design or previous measurements.

In principle, all the major energy users, such as chillers, cooling towers, boilers, and large motors, should be separately monitored. Under normal operating conditions, this allows specific problems to be highlighted. When energy conservation measures are being implemented, it allows for detailed feedback on their effectiveness.

The International Performance Measurement and Verification Protocol (see www.evo-world.org), which was partially funded by the U.S. Department of Energy, developed recommendations for equipment monitoring to help manage energy use and to provide verification when a building management team is using an external energy performance contract to manage energy reductions. This protocol is widely used in the United States and some other countries,[11] and it is the base standard for the monitoring credit in LEED.

There are two pieces of additional information needed to allow accurate evaluation of measurement and verification data against a baseline: weather data for the time periods being considered and the number of hours for which equipment is running (e.g., systems left on to cover out-of-hours working). These are the two primary factors that change energy use in an existing building that are not subject to management by the facilities team.

Energy Conservation Measures

Once a building owner has identified that there is potential for energy savings, the next step is usually to identify the specific changes that might achieve those energy savings. These changes are called energy conservation measures (ECMs) or energy efficiency measures (EEMs).

As for new buildings, the priorities in an ECM program are to reduce energy demand and operating hours (see also Chapter 7). Some equipment efficiency improvements or controls upgrades may also be economically feasible.

Top Ten Residential Energy Conservation Measures

1. Turn thermostats up in summer and down in winter
2. Turn systems off when not in use
3. Draw shades/shutters to reduce summer heat gain
4. Air-dry clothes
5. Retrofit lighting
6. Use Energy Star appliances (TV, refrigerator, washing machine, dryer)
7. Use an energy monitor
8. Insulate the building
9. Use an energy-efficient boiler and air conditioner
10. Reduce air leakage of walls and windows

In single-family homes and other small buildings, where there is likely to be no analysis or consultant budget, the best approach is to use a top-ten list of ECMs and select those items that clearly apply. Payback calculations are either unnecessary or often provided by the equipment manufacturer, installer, or utility.

In more complex buildings, analysis is usually required: the money required to make changes is greater, and there are many more people involved in approving spending.

There are two main places to look for savings in existing buildings: operations and efficiency. Often, complex buildings are not operating as they were designed to work. When automatic controls systems were introduced in the mid-1980s, designers started to include energy-saving protocols, such as optimum start and modulation of fresh air supply volumes depending on outside conditions. As electronics have become cheaper, ever more complex energy saving strategies have become possible and have been adopted by design engineers. Many of these strategies save significant energy when operated properly and have become routine, or even required by code, but the automatic nature of the controls means that no one has to watch what's going on closely. It has not always been recognized that the systems have to be set up and tested with care—every system is different, and there are interactions between the building and the system that are very difficult to predict. The setup process is called commissioning, often abbreviated to Cx, and it occurs at the end of the construction process, when the schedule is often tight and money is in short supply, so it is not always perfectly executed.

In addition, if a building has been in use for some years, the sensors may have slipped out of calibration or adjustments may have been made for specific events or conditions—for example, extending operating hours or changing a temperature set point—that were never corrected. Examining and correcting the building operations is a process called retrocommissioning, or RCx.

To develop ECMs for a particular building, the first step is to understand the building systems and perform an energy audit. Energy audits can vary widely—ASHRAE defines three levels. The most basic level of audit (Level 1) is a walk-through analysis and generates a list of low-cost ECMs based on the walk-through. A Level 2 audit involves a walk-through, an energy survey, and an outline engineering analysis, and results in a list of potential ECMs that have been prioritized based on likely cost. A Level 3 audit includes detailed energy modeling of ECMs to calculate payback periods for all ECMs, including those that are capital intensive.

Top Ten Commercial Energy Conservation Measures

1. Monitor energy use and report it up the hierarchy
2. Turn systems off when not in use
3. Eliminate waste—steam-trap leaks, leaking valves or dampers.
4. Retrofit lighting
5. Change nighttime setback temperatures and implement optimum start
6. Control amount of fresh air in the system for energy efficiency (sometimes more, sometimes less)
7. Change to variable flow systems
8. Turn thermostats up in summer and down in winter
9. Use premium-efficiency motors
10. Replace equipment with more efficient equipment

The list of ECMs should be tailored to the building. However, there are a number of strategies that apply to almost every building and can be shown to have very rapid paybacks. Examples of these types of strategies are in the "Top Ten Commercial Energy Conservation Measures" box. The order will vary depending on current conditions and systems.

The more complex the systems and the poorer the energy performance, the more likely it is that a full ECM analysis and retro-commissioning approach will be needed. The ECMs may rely on the existing equipment or may need the addition of new sensors and controls. The specific approach and results will vary building by building.

Occupant Behavior

Occupant behavior can change energy use enormously. Occupants can open or close windows, adjust the settings on the local thermostat, determine whether lights and computers are switched off, use local fans and heaters, and make complaints. In office buildings, tenants can change the energy use by between 30 and 80 percent, depending on the climate.[12]

In residential buildings, the occupants usually pay the energy bill. It is therefore often enough to educate the occupant about the ways in which they can save energy, and it will happen. The chances of this working can be increased by providing a monitor that provides instant feedback about the rate of current energy use.

In commercial buildings, there are usually many occupants, and very few of them are concerned with the energy use of the building. Their priorities have to do with, or should be, getting their work done, and the attitude is often "Make me comfortable, or else!" The energy use is usually managed by a building operator or facilities manager who has no direct interface with the occupants except to receive hot/cold complaints. Changing occupant behavior is therefore about both education and motivation.

In some commercial buildings, there is very little that the occupants can adjust. In North America, it is rare to have operable windows because they are expensive and introduce another complication into energy management. Thermostats are often shared between several people, so changing the setting can be a complex negotiation. People's seats are also often fixed, so they cannot move to avoid sun

UCSF Mount Zion Energy Saving Case Study

The University of California, San Francisco (UCSF) Mount Zion Research Center is a four-story structure built in 1997–8; the building's primary function is as a cancer research laboratory. Mount Zion had the highest energy consumption of all the research centers at UCSF, with 37 percent greater energy intensity than the next highest energy consumer. In 2007–8 UCSF hired a team to help create significant energy savings in existing buildings through retrocommissioning and effective metering and monitoring.

The project was carried out in three stages. Firstly, an initial opportunity assessment for monitoring-based commissioning and retrocommissioning at the Mount Zion research center was carried out. This established baseline data for energy use and costs and then identified opportunities that would lead to a significant reduction in energy consumption. Following this assessment the team was appointed to provide design, project management, and commissioning and verification services to realize the opportunities identified in the first-stage assessment. After implementation, a report was issued to show the technical implementation and energy saving results for the project.

The initial study showed that the primary cause of unnecessary energy consumption was excessive airflow due to a lack of airflow monitoring, which in turn affected the peak cooling and heating loads and caused the zone reheat coils to be in a near-constant state of heating. To solve this, daytime and nighttime airflow rates were reduced. This generated energy savings by reducing fan power requirements and the need for heat, which in turn reduced boiler natural gas consumption. Further savings were generated by addressing air balancing issues and resetting unoccupied space temperature set points in noncritical building spaces such as offices and corridors.

Before the project, UCSF had no way to monitor energy usage and was therefore unable to identify areas of intensive energy use. The project has assisted UCSF to resolve this problem and also provided cost-effective energy savings. A number of other savings based on the improved building monitoring capabilities may be realized in the future.

9.8 UCSF Mount Zion Research Center
Image: Arup

TABLE 9.1
Energy and cost savings compared to business as usual
Graphic: Arup

	Electricity (kWh)	Natural gas (therms)	Total (MMBtu)	Electricity costs	Gas costs	Total
Baseline energy	5,369,087	397,569	58,082	$691,799	$381,666	$1,073,465
Post energy	3,367,236	297,075	41,200	$433,700	$285,192	$718,892
Savings	37.3%	25.3%	29.1%	$258,099	$96,474	$354,573

(a)

9.9 "Take the Stairs!" poster from NYC
Image: www.nyc.gov

9.10 Operable window details
at (a) Selly Oak College, Birmingham,
UK, and (b) Kroon Hall, Yale School
of Forestry & Environmental Studies;
note the lights on the Kroon Hall
image—these are used to show when
windows can be opened and when
they should be closed
Images: (a) Arup, (b) Morley von Sternberg

or drafts. All of these factors tend to cause occupants to think that they have no influence over energy use. However, there is always something they can do, from turning off lights and computers to taking the stairs, closing the blinds on east-facing windows in summer before they leave for the day, and accepting that the office might be warmer in summer than in winter. In Europe, operable windows and mixed-mode buildings are more common, as are automatic shades, but the sense of having no control is similar. Some examples of how operable windows can work in commercial buildings are shown in Figure 9.10.

(b)

As in residential buildings, energy awareness is a very good way to reduce carbon emissions—it raises awareness of the issues among the staff, it can make a real difference to energy use, and the only money it usually requires is for training. One of the leaders in building dashboards, Lucid Design, has shown measurable savings in most of the buildings that have its system installed.[13]

BARRIERS

If the opportunity is so great, why aren't carbon emissions reductions projects more common? There are many reasons why retrofit projects are hard to start.

Taking the project steps in turn—the people managing buildings often don't know that they have a problem. Even if they do know, they rarely have the right combination of funding, information, expertise, and tools available to be able to reduce energy use without hiring or enlisting help from both inside and outside their organization.

There are likely to be some elements that can be dealt with without consulting help, perhaps including obvious maintenance issues like leaking steam traps or overlong operating hours. These can often be solved within regular maintenance budgets with the available skills, and the energy bills will reduce, but these "free" improvements will eventually run out, and achieving the best possible energy use will need some investment. It is also possible that the extra thinking needed from the maintenance team will not be rewarded or even encouraged. If outside help is needed, a budget is needed, which means that there has to be a justification for the spending. Sometimes there are no clear existing system drawings and no description of design intent, which means that the work to identify likely strategies and paybacks for a conceptual business case can be hard to do. The need for a business case alone can defeat some projects if the expertise to produce one is not available within the maintenance team.

If a list of ECMs, a budget, and a business case are assembled, the next barrier is to find the funding for the upgrades. This usually means that a loan will be needed. Building owners are always reluctant to take out an additional loan against their asset, even if it can be shown that the loan will be paid back from energy savings, because it appears to reduce the value of their property. Low long-term energy use should increase the value of a property, but this has historically

Harvard's Green Loan Fund

Harvard's Green Loan Fund is a $12 million revolving fund that provides capital for resource reduction measures in Harvard's buildings. Loans are limited to $500,000 per measure and are paid back with money saved on energy, water, or waste management costs. Loans can be applied to new and existing buildings and can be for either the full cost of an activity or the additional cost associated with reducing the environmental impact. Full-cost loans must have a payback of less than five years and incremental-cost loans have to have an internal rate of return of 9 percent or better. Funding has been provided for nearly two hundred projects at a cost of $15 million, resulting in nearly $5 million in savings, and with a median return on investment of nearly 30 percent.

been a small factor in appraisals and how this should be reflected in valuations is still subject to argument.

It would be possible to finance some equipment upgrades from operating budgets if the operating and energy budgets were "ring-fenced" year on year, such that savings could be reinvested in operations. For example, a lighting retrofit and a focus on reducing unnecessary operating hours might pay for themselves within one year, yielding a savings in the second year that could be spent on premium efficiency motors, which would yield a further saving that could be spent on monitoring equipment. This almost never happens: if there are energy cost savings, these are immediately returned to the center and taken as profit rather than held in the operations budget. Systems like Harvard's Green Loan Fund (see box) can help hold energy savings for energy reduction activities.

There are variations in the paybacks of different ECMs, and there is always a tendency for financial managers to pick only those that pay back very quickly. In the absence of a system that allows reinvestment of energy savings dollars into energy upgrades, this means that once the low-hanging fruit has been picked, it becomes harder and harder to justify energy upgrades. To achieve the best long-term results, it is usually necessary to group ECMs and calculate a combined payback period. Sometimes the best ECMs will be visible only to energy specialists.

Corporations will compare their investments in energy efficiency against other investments that they might make. They will often define the minimum acceptable rate of return for investments and this is known as the *hurdle rate*. Different types of building owners will have different maximum paybacks or hurdle rates. Typically, a corporation will be looking for a two- or three-year payback while government and public sector owners may be comfortable with a 10-year payback. This is one reason why bundling ECMs may lead to a better long-term result: the minimum hurdle rate can be achieved by combining ECMs with very short paybacks with longer-term ones.

In some commercial leasing markets there also may be what is described as a *split incentive*. This occurs when the person who pays for equipment upgrades does not get the benefit of the savings. An example would be when a tenant pays the energy bills and uses cooling from a common chiller: if the owner pays for a chiller upgrade, only the tenant will see the benefit. A similar problem can occur if tenants pay a flat rate for energy to the building owner, because they will not be incentivized to turn off equipment when it is not in use or otherwise to manage their energy use. There are many ways in which lease structures can be a barrier to carbon emissions reductions through energy saving.

Another financial barrier to carbon emissions reductions projects is that paybacks for significant changes can be quite long, in the order of seven to twenty years. This issue is particularly acute in North America, as a result of low energy prices and lack of carbon policy. Many building owners do not know if they will still own a particular asset at the end of that time, so they are not inclined to invest when they will not realize the benefits themselves and when it is not clear how the energy improvements will be valued by the appraiser. If they have to borrow money to make the upgrade and already have a loan on the property then this second loan complicates the terms of the first, or may not be allowed at all.

The final barrier to the implementation of retrofit projects is the presence of tenants in commercial buildings. This can make it difficult to get access to every space without disturbing their activities. There is therefore a need to coordinate upgrade activities with the tenant turnover cycle.

POLICY AND FUNDING MECHANISMS

A recent study by the World Economic Forum shows that there are two key means by which retrofits in existing buildings can be encouraged. The first is through policy and the second is through access to capital to do the work.[14]

The reason why policy is so important is that it provides a driver for building owners to make positive changes that would not necessarily be discovered otherwise, or if discovered would not be considered to be priorities. Policies take various forms: in some jurisdictions there is simply a requirement to report on energy use, with the intention that the owners of poorly performing buildings will then be aware of shortcomings and make corrections. A more robust policy intervention requires building energy audits that identify specific ECMs with their payback periods. Alternatives to these studies include mandating simple changes such as lighting retrofits. Policy can also be used to provide funding to improve payback periods. Typical mechanisms include part-funding for energy studies and audits or subsidies for energy efficient equipment. The funding for these policies can be provided from a surcharge on energy bills or through tax relief. Robust, mature policy is likely to address all of the elements of benchmarks, baselines, efficiency improvements, and funding mechanisms. These policy mechanisms can apply to both residential and commercial buildings.

Funding is often problematic, as noted previously, because it will require an additional loan on a property. Even where funding is possible there are barriers: valuations do not always recognize energy efficiency improvements clearly and owners do not always want to put the cost of improvements on the balance sheet. Loans specifically for energy efficiency are also fairly new, and there is no clear way to treat them within banks. In the last couple of years there has been increased interest in generating "green loan funds" that focus on energy retrofits. These can be difficult to put together because of the scale of projects: a typical retrofit project might cost $100,000 to $1,000,000, so that an investable fund needs to aggregate hundreds or thousands of projects from an industry that is notoriously segmented.

There are two mechanisms for funding energy retrofits that change the basis of the loan. The first of these is the Property Assessed Clean Energy (PACE) loan. PACE loans focus on renewables retrofits and allow people to borrow money from the municipality, paying it back as an additional property tax over time. The advantages are that loans can be made for a limited menu of options, reducing the need for expensive analysis, and the loan remains with the property so that property owners are free to move without having to continue to pay off the loan. Many states passed legislation to allow this to happen but uptake has been limited because taxes take precedence over mortgages, so many lenders have not allowed their borrowers to take advantage of the scheme.

The second mechanism is MESA (Managed Energy Service Agreements), developed by Transcend Equity Development. Under this model a third party takes the loan for the property improvement and assumes responsibility for paying all the energy bills. This takes the loan off the property's balance sheet, and the loan is repaid by the third party from savings in the energy bills.[15] Clearly, this model requires up-front analysis and is only applicable to large commercial buildings.

NOTES

1 New York: Penguin, 1995.

2 Brookings Institution.

3 J. Heller, M. Heater, and M. Frankel, "Sensitivity Analysis: Comparing the Impact of Design, Operation, and Tenant Behavior on Building Energy Performance," white paper, Vancouver, WA: New Buildings Institute, 2011, http://newbuildings.org/sites/default/files/NBISensitivityReport. pdf (accessed June 10, 2012).

4 "IFMA Survey Ranks Top 10 Office Complaints… and Some That Score High on the Laugh Meter," BUILDINGS, December 23, 2003, http://www. buildings.com/tabid/3334/ArticleID/1689/Default.aspx (accessed June 26, 2012). The top 10 are (1) too cold, (2) too hot, (3) poor janitorial service, (4) not enough conference rooms, (5) not enough storage/filing space in workstation, (6) poor indoor air quality, (7) no privacy in workstation/ office, (8) inadequate parking, (9) computer problems, and (10) noise level/too noisy.

5 Australian Government, "Commercial Building Disclosure," http://cbd. gov.au/default.aspx?test=true (accessed June 26, 2012).

6 New York City's Local Law 84 of 2009 is one example: http://www.nyc.gov/ html/dob/downloads/pdf/ll84of2009.pdf (accessed June 10, 2012).

7 Brandeis, University of British Columbia, and UCLA have all had "Do It in the Dark" energy challenges for their residential halls.

8 National Grid ran a pilot amongst 50,000 Boston, Massachusetts–area customers showing how energy bills compared to neighbors', directly on the monthly invoice, starting in 2009. Energy reductions of about 2 to 3 percent typically result. See E. Ailworth, "Energy Report Card Aims to Boost Conservation," The Boston Globe, October 9, 2009, http://www. boston.com/news/science/articles/2009/10/09/national_grids_energy_ report_card_aims_to_boost_conservation/ (accessed June 10, 2012); J. Schmit, "Do You Use More Energy Than Your Neighbors?" USA Today, February 1, 2010, http://www.usatoday.com/money/industries/energy/2010- 02-01-homeenergy01_ST_N.htm (accessed June 10, 2012).

9 New York State Public Service Commission.

10 Chartwell, "Hydro One Offers Free Electricity Monitors to 30,000 Customers," Chartwell's Best Practices for Utilities & Energy Companies, 9(1), January 2007, http://sites.energetics.com/MADRI/pdfs/ChartwellHydroOne MonitoringProgram.pdf (accessed June 10, 2012).

11 France, Romania, Bulgaria, Poland, Spain, Czech Republic, and Croatia have their own versions of the protocol.

12 Heller, Heater, and Frankel, 2011 (see note 3 above).

13 See http://www.luciddesigngroup.com/presskit.php.

14 World Economic Forum Retrofit Finance & Investing Project, "A Profitable and Resource Efficient Future: Catalysing Retrofit Finance and Investing in Commercial Real Estate," Cologny, Geneva, Switzerland: World Economic Forum, October 2011.

15 See www.transcended.com.

i See http://www.rmi.org/retrofit_depot.

ii "Pathways to a Low Carbon Economy: Version 2 of the Greenhouse Gas Abatement Cost Curve," January 2009.

Kroon Hall, Yale School of Forestry & Environmental Studies integration
of structure, daylighting, solar shade and renewable power
Image: Arup

10

GUIDELINE CHAPTER

Integrated Design

Alisdair McGregor

The term "Total Architecture" implies that all relevant design decisions have been considered together and have been integrated into a whole by a well-organized team empowered to fix priorities.

Sir Ove Arup, 1970

Up until the start of the twentieth century, all building design had to use a holistic design approach. Buildings had no mechanical and electrical systems to compensate for ineffective building design. Essentially, all buildings were climate responsive. In general, building materials were local as were the building trade skills. The structural engineer had to understand the limitations of the local materials and trade skills. Architects had to understand the climate and the available building resources. In other words, an integrated approach was essential. The availability of electric power and the almost universal use of air-conditioning have enabled the design of the building to be uncoupled from the thermal and lighting characteristics of the design. Modern materials and computer power have enabled the structural engineer to design dramatic structures, but all too often that skill is used to solve problems that could have been avoided by a creative team working together from the concept phase.

At the time that this "technology can solve all our problems" approach was developing, the professions were fragmenting into specialists. The engineering professions, perhaps in response to fee competition and fear of litigation, narrowed their focus to solving the problems of their particular discipline with little regard to the overall performance of the project. Add to this mix the typical contractual relationships between owner, architect, engineer, and contractors—which seem to rely on setting up adversarial relationships between the various parties—and we begin to see why setting up an integrated design and construction team is not as simple as it may seem.

It was frustration with this fragmentation that led Sir Ove Arup to define what he called *Total Architecture*, back in 1970. What he describes is the essence of integrated design.

A common trait in an integrated design is the use of one system or component to serve multiple functions. One of the most striking examples of this multifunctional integration is the Menil Collection gallery in Houston by Renzo Piano (see the case study in the box overleaf). If you can't see the joints between the different design disciplines, then it is an integrated design.

There are times when full integration is not the best solution. When some building systems are likely to require regular modification it is best to decouple these systems from other components. For example heating, ventilation, and air-conditioning systems in hospitals need to be modified on a regular basis as new medical procedures demand functional changes. So for hospitals, the mechanical and electrical systems are separated from the long-term components, such as the structure, and are arranged so as to be easily accessible. Another example is radiant cooling/heating pipes cast into the concrete structure, which are a very efficient space-conditioning system and a great example of integrated design. However, the long-term flexibility of the system is limited. The solution is to design for an easily adaptable task-conditioning system to provide flexibility and keep the efficient integrated radiant system for the base load. The ideal design will have a mix of tightly integrated components and deliberately decoupled systems.

WHY IS INTEGRATED DESIGN IMPORTANT?

Integrated design is not easy. It requires team members to respect each other and work for overall success rather than firm or individual success. During the early phase of design, it can seem to circle around, which can frustrate the "ready, aim, fire" mentality of project managers. However, the rewards are buildings or developments that are balanced and harmonious, meet the energy performance goals needed to mitigate climate change, and cost the same or less than traditional projects.

Integrated design is very much a buzzword in the design and construction industry. Unfortunately, as with *sustainable design*, it has multiple definitions. It is often assumed to be synonymous with coordinated design. However, though all designs should be coordinated not all designs are or even should be integrated. Integration must start at the very beginning of a project, whereas coordination can occur much later. It is entirely possible, but not recommended, to have a poorly integrated design that is well coordinated.

Building information modeling (BIM) is not integrated design but certainly helps both coordination and integration. Integrated

ALISDAIR MCGREGOR

Case Study: The Menil Collection, Houston, Texas

The finished gallery spaces are elegant and simple in their design, allowing art to be seen in optimized conditions. The simplicity belies the hours of design workshops, testing, and collaboration with manufacturers and builders.

The "blades" shown in Figures 10.1 and 10.2 bring reflected light into the gallery space without any direct light, which is harmful to art. The conditioned air is delivered at floor level as a displacement ventilation system. This is not only energy efficient, it avoids cluttering the ceiling space with supply ducts. Exhaust air is removed above the blades so that solar gain through the glazed roof is trapped between the glazing and the roof, and removed before it enters the conditioned space. A single element functions as architecture, structure, mechanical system, and daylighting device.

10.1 The Menil Collection, Houston, Texas; the ferro-cement blades are structure, daylight control, return air path, and architecture
Image: Richard Bryant/arcaidimages.com, Architect: Renzo Piano Building Workshop

10.2 The blades also form the bottom cord of the roof truss; Arup worked closely with a concrete-boat builder to develop the construction technique for precasting the ferro-cement blades
Graphic: Arup

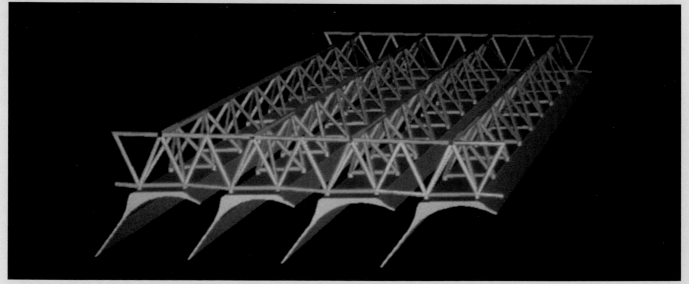

project delivery (IPD) is not integrated design but most definitely helps the integration and coordination process, if all parties have equal input to the design concepts. Unfortunately, IPD can devolve into a traditional linear design process, though with everyone in the same room.

An integrated design requires the whole design to be optimized, balancing the different components to produce a harmonious whole. This is in contrast to buildings where the mechanical systems are overdesigned to compensate for a poorly designed envelope, or the structure uses excessive material to compensate for architectural whim. An integrated approach deliberately looks for synergies between components so that one component can serve multiple functions. Done creatively, this can result in cost savings, energy savings, and reduced space for building systems.

The integrated process does require that all the players in the design process are involved from the outset. This means that more time is required in the early design phase, something that has to be recognized in the fee structure. The uptake of BIM requires a greater participation of the whole design team up front, so the design effort moves to an earlier stage. The integrated design process can appear chaotic at times, and there will be many ideas generated that go nowhere. However, the result is a composite design that works as a whole. In comparison, the traditional linear process can seem more ordered and controlled, but runs the risk of baking in inefficiencies at each step.

HOW DO YOU INTEGRATE DESIGN?

Although the integrated process can appear chaotic, there does in fact need to be organization to get the best results. Merely putting all the design and construction players in the same room will not necessarily produce results. It is essential to understand which components require integration and who the key members who input to those components are. This avoids having team members who have little input to the particular topic feeling unproductive through a long design meeting.

At the start of a project, general workshops are required to set the goals for the project. These workshops require the entire team, including the owner and contractor if they are already engaged. The goal is to set a common purpose and agree on what success looks like. By the end of the workshop, participants should be talking about "our building" and what "we" are going to be doing to make it a success. Integrated design requires egos to be left behind. There is no place for people who are interested in just their component of the design.

The key components for success are as follows:

- Multidiscipline teams work together from the start of the project
- The whole building is optimized, not just the components
- The whole team has contact with the owner
- The contractor is involved early, providing constructability advice.

The first two components concern the approach to design. Getting low-energy performance with minimal extra cost requires optimizing the building before the systems are selected and designed. Similarly, structural systems can be made far more efficient if the structural considerations are built into the design from the earliest concepts. Examples include optimizing the column grid for the floor plan and future flexibility, selecting a lateral system, and choosing the floor construction considering both function and cost. The structure can be designed to be part of other building systems, such as the thermal storage system or shading system for the envelope.

The use of an integration diagram can help determine where different building components or strategies interact with each other. The diagram can aid in deciding who needs to be present for a particular part of the design. Figure 10.3 (overleaf) shows a project where the use of natural ventilation and daylight are to be maximized. The components and strategies of natural ventilation and daylighting that need to be integrated can be identified. These two strategies would normally be the responsibility of the mechanical engineer and the lighting designer, respectively. By diagramming the integration points, the other team members needed for a well-integrated solution can be identified.

10.3 Integration diagram plotting all the interactions that need to occur to integrate daylight and natural ventilation in a building
Graphic: Arup

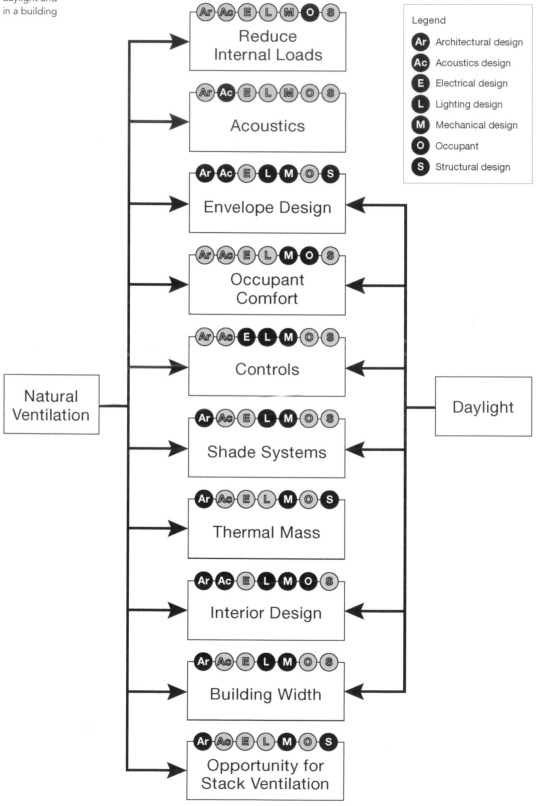

Legend

- Ar Architectural design
- Ac Acoustics design
- E Electrical design
- L Lighting design
- M Mechanical design
- O Occupant
- S Structural design

Reduce Internal Loads

Acoustics

Envelope Design

Occupant Comfort

Controls

Shade Systems

Thermal Mass

Interior Design

Building Width

Opportunity for Stack Ventilation

Natural Ventilation

Daylight

Another aspect of successful integration is looking for and identifying potential synergies where one design element can enhance another or even serve multiple purposes. Building-integrated photovoltaics are a good example.[1] In most cases, single-purpose photovoltaic installations do not have a good economic payback. However, by making them act as a shade structure, say over rooftop air-conditioning units, not only are they providing a visual screen, they are reducing the air temperature at the intake to the air-handling units, thereby reducing the energy consumption of the building while generating renewable energy. When the cost of items that are removed (rooftop equipment screens) is subtracted from the economic equation, photovoltaics can be shown to be a good investment.

The envelope is perhaps the key component that demands an integrated design approach in a high-performance building. The envelope is part of the lighting and mechanical system, and possibly part of the electrical generation system; it affects the well-being of the occupants; it sets the aesthetic of the building; it keeps the weather out; and it has to meet structural performance standards. Clearly, a lot of people need to be involved right from the start of the envelope design. If done correctly, an optimized envelope not only saves energy, it can also reduce or eliminate cost in other systems.

This brings us to another essential team member: the cost estimator. Compartmentalized cost estimating can ruin the best integrated design. Clearly, a holistic design requires a holistic cost estimate. Take the example above of the integrated envelope. The high-performance envelope may exceed the "allowance" for the envelope in the budget. However, the size of the air-conditioning system may have been reduced, or simplified if solar gain is reduced, or possibly the air-conditioning can be eliminated. This approach of pushing a design element to extremes so that whole systems can be eliminated has been advocated by Amory Lovins as "tunneling through the cost barrier."[2] The case study of the Jerry Yang and Akiko Yamazaki Environment and Energy Building (see box) illustrates this method. Initial concept cost estimates showed a possible premium for the LEED Platinum design of just under 10 percent. After the glazing and atrium design had been optimized, it was determined that no air-conditioning was needed for the north- and east-facing faculty offices. The net result was a cost premium of less than 2 percent.

Case Study:
Jerry Yang and Akiko Yamazaki Environment and Energy Building

The four atrium spaces in the Jerry Yang and Akiko Yamazaki Environment and Energy Building (Y2E2) at Stanford University were part of the organizing structure providing casual meeting spaces and encouraging interaction between the different research groups in the building. Measuring 81 feet from basement to highest point, all four atria extend the full height of the building and allow the integrated disciplines to collaborate and connect, be inspired and curious, and remain transparent and open to daylight, view, and outside air.

The extensive use of glass at all levels and the design's sensitivity to sight lines reveal laboratory work, flexible seminar classes, administrative activities, and social gatherings, making all the building occupants visible to visitors and, most importantly, visible to each other. At the same time, the design team realized that the atria could bring daylight into the interior spaces and act as a driver for natural ventilation.

The final solution required the close collaboration of architects, mechanical and electrical engineers, lighting designers, acousticians, fire engineers, contractors, and cost estimators. There were numerous collaborative design meetings and over sixty computational models were created, reflecting over 2,500 hours of computer analysis. This integrated design effort resulted in atria that work on multiple levels. The users experience light, airy spaces without the acoustic reverberation often found in such spaces, unaware of the hours of collaboration that went into optimizing the solution.

Ventilation
The atria act as the building's lungs, passively drawing air through natural buoyancy and cross-driven pressures up and out of the louvers near the apex. Evenly positioned throughout the building, they are integrated into the layout, allowing direct connection and ease of natural ventilation for all the perimeter offices on every facade.

Given the potential strength of airflow through all the atria openings, the team needed a strategy to manage the velocity while maintaining enough passive draw during warm, windless conditions to move adequate air through the building. The result was a combination of motorized and manual windows combined with motorized louvers. Through the building management system, the motorized windows open and close when interior

continued overleaf

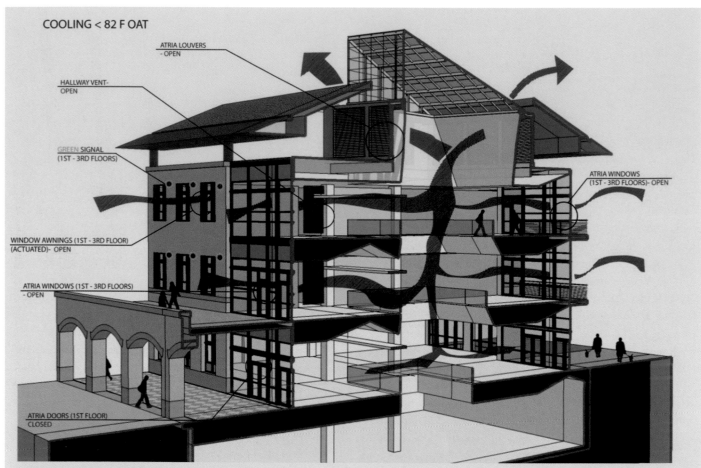

COOLING < 82 F OAT

ATRIA LOUVERS
- OPEN

HALLWAY VENT-
OPEN

GREEN SIGNAL
(1ST - 3RD FLOORS)

ATRIA WINDOWS
(1ST - 3RD FLOORS)- OPEN

WINDOW AWNINGS (1ST - 3RD FLOOR)
(ACTUATED)- OPEN

ATRIA WINDOWS (1ST - 3RD FLOORS)
- OPEN

ATRIA DOORS (1ST FLOOR)
CLOSED

10.4 The natural ventilation system through the atria
Graphic: Arup

temperatures reach a preset point relative to outdoor temperature, providing fresh air to all levels of the atria as well as cooling the corridors. Simultaneously, manually operable windows and ceiling fans provide and circulate fresh air to and through offices at occupant discretion. The release for all air entering the building in both ways is in the glass atria caps, which have louvers on all four sides. A rooftop wind speed and direction monitor tells the building which louver dampers need to open and close. In windy conditions, louvers downwind of the atrium cap open, creating negative pressure to draw air out of the building, which in turn creates more draw to pull more air out. When there is no wind, all louvers open. The atria louvers modulate for finer control.

Fire Engineering
California fire safety codes require smoke control in atria interconnecting three or more stories. Instead of providing the usual mechanical exhaust systems, however, Arup's fire engineers and Stanford chose to capitalize on the natural ventilation design,

which would allow openings between the first floor and the basement to supply natural light and visual connection to the lowest level.

After preliminary calculations, the team gave the architect and the mechanical engineers an estimate of ventilation opening sizes, which was then used in design development. Computational fluid dynamic modeling enabled the fire team to determine the smoke development and tenability of the space, as well as the benefits of venting by passive means.

There was concern that a basement fire may not have sufficient buoyancy to vent passively. The two conventional solutions—providing more openings at low levels or closing off the basement—did not match the architectural goals of the atrium spaces. The adopted solution was to provide a horizontal fire shutter that closes off the basement portion of the atrium in the event of a fire.

As part of the commissioning process, the authorities required live hot-smoke testing, the success of which surprised even the fire marshal.

Daylighting

Another objective of the atria was to bring daylight into the core of the building. Conventionally, the openings at each floor would be uniform. However, in order to provide light without glare or thermal discomfort, the lighting team worked with the architect to optimize the design by modifying the openings. Three of the atrium openings were refined to be widest near the top of the building and narrowest at the base. This allows the floor at each level to extend a little below the opening in the floor above, with more floor space at the south sides of each atrium. These receive abundant ambient light, as direct solar light primarily arrives at the north. The south sides are thus ideal locations for meeting rooms with glass walls, open to ambient light and visually connected to the building's activities. Daylight is also available to interior work spaces from the atria. Occupants report that they rarely resort to the architectural lighting during the day, preferring to have either no lights or low-energy LED task lights.

Acoustics

For the atria interiors, room acoustic requirements were again bound by openness and transparency requirements, effectively establishing "go/no-go" zones for provision of acoustic treatment. Acoustic treatment in the atria was needed to reduce reverberation time and diminish occupancy noise and break-in noise from the building services systems.

10.5 Open work space with daylight
Image: Tim Griffith, Architect: Boora

Preliminary studies indicated that without acoustic treatment the rooms would be highly reverberant. Opportunities for treatment were, however, limited, due to the lack of available wall space and restrictions on putting acoustic banners at the tops of the atria (blocking light), along the sides of the upper atrium cupolas (blocking air flow at the louvers), or on the floor (more difficult to keep clean than hard floors). The design team built an acoustic

continued overleaf

10.6 The Jerry Yang and Akiko Yamazaki Environment and Energy Building
Image: Arup, Architect: Boora

10.7 A basement area in the core of the building shows the penetration of daylight
Image: Tim Griffith, Architect: Boora

model of a "typical" atrium (all four are slightly different) and placed sound "sources" at various locations within. Arup's acoustic engineers wanted to determine, for a given source location, which surfaces in the room were acoustically "open." By analogy, if the acoustic source was represented by lighting, the team wanted to see which surfaces would be "illuminated" and which would remain dark. Those surfaces illuminated by a particular source location would be the first choice for placing acoustic treatment to control the direct and early reflected energy from that source. By absorbing and reducing these sound energies efficiently,

control of the reverberant energy was also optimized. Repeating this for sources at each floor level adjacent to the atria and at the atrium floor levels themselves yielded a list of surfaces that were prime locations for acoustic control of reverberation and occupant-generated noise. By vetting the effectiveness of the various room surfaces for acoustic accessibility to the direct and early sound energy from the sources, Arup could work with the architect to prioritize and focus on a specific palette of surfaces for treatment.

HOW DO YOU KNOW WHEN YOU ARE DOING IT?

If you're not having fun, then it probably isn't integrated design. The teamwork for successful integration requires a level of cooperation and trust between team members that goes beyond the contractual requirements. The final product should be a fully optimized solution where the integration of the different disciplines is seamless.

NOTES

1   Simon Roberts and Nicolo Guariento, *Building Integrated Photovoltaics: A Handbook*, Basel, Switzerland: Birkhauser, 2009.

2   Paul Hawken, Amory Lovins, and L. Hunter Lovins, *Natural Capitalism*, New York: Little, Brown and Company, 1999.

Arup grants over $8 million per year to staff for research and development activities.
This blue sphere represents the community connections between staff billing time to
R&D grant numbers and suggests the diverse emergent knowledge that is developed
through evaluation
Image: Arup/Duncan Wilson

11

How We Choose
Evaluating Strategies and Trade-offs

Cole Roberts

What's the difference between an accountant and a climate change scientist? An accountant is precisely wrong. A climate change scientist is approximately correct.

Source unknown

DECISION MAKING AND THE SUCCESS OF CIVILIZATIONS

All great civilizations make decisions. Those that succeed make the right ones. Pulitzer Prize–winning author Jared Diamond researched the role that choice played in either the collapse or resilience of societies.[1] He concluded that crucial to success was not only action, it was *right* action. But how do we choose the right action? Through understanding what type of decision making dominates our choices, why we make the decisions we do, and how to make them better, we move toward successful response to climate change.

As discussed in Chapter 10, a significant opportunity for better decision making occurs through integrated design and planning – the developing of options that solve multiple challenges through singular action. The results of such integrated effort go far beyond the incremental gains of efficiency because they capture the *synergistic* benefits that exist at the intersection of multiple expertises. But developing integrated options isn't enough. We must take into account the one human condition that defines our past, present, and future decision making, the *default condition*.

Here is Edward Bear, coming downstairs now, bump, bump, bump, on the back of his head, behind Christopher Robin. It is, as far as he knows, the only way of coming downstairs, but sometimes he feels that there really is another way, if only he could stop bumping for a moment and think of it.

A. A. Milne, *Winnie-the-Pooh*[2]

THE DELUSION OF RATIONAL OUTCOMES

Rational choice is oversold and under-realized. We are predisposed through education and upbringing to believe that reasoned argument will result in reasoned decision making. We believe that if knowledge is input, the *right* answer will churn through the decision-making apparatus and fall at our feet on the other side.

Irrational choice is often the stronger voice, for ourselves, our collaborators, and our clients. People routinely make decisions based on partial information, inaccurate information, prejudgment, postrationalization, or too hastily to develop thorough understanding. Even on the rare occasions when adequate information, expertise, and time are available, emotions are often the trump card. "How can this hurt me?" "This will be the first time, how exciting!" "No one gets sued for conservatism." "Award-winning for sure!" "I know it's right, but I can't support it." These are just a few examples of statements, whether spoken or not, that are underpinned by emotions like fear, belonging, selfishness, baseless distrust, and blind optimism.

11.1 There really is another way
Image: Copyright of the Estate of E. H. Shepard

Yet the choice that defines the direction of our built environment, more than reasoned argument or irrational exuberance, is elegantly and simply what came before: the *default condition*.

The default condition is safe since others did it (think protection in groups). It is easy since we've done it before (think existing tools). Its outcome is known since we can see it (think existing data). It is inexpensive since anything better or new should always cost more (think marketing). It is hard to change (think existing city streets). It is politically nonconfrontational (think NIMBYism[3]). It is appropriate since it reflects our culture (think the sexy automobile). It is cost effective since the financial system knows how to pay for it (think loan underwriting).

The default condition is an odd boogeyman since no one creates it, or directs it, or defines it. It emerges from us all—a collective creeping normalcy and landscape amnesia that makes it hard to internalize the past for the sake of the future.

The default condition is the single greatest hurdle to overcome in our efforts to improve the built environment. It is also the single greatest opportunity. If the default condition can be *reset,* everything that is better about the *reset* default condition is now just as easy, safe, inexpensive, and right.

So if rational choice (say, one-third of the time) and irrational choice (the other two-thirds of the time) can lead to the successful resetting of the default condition, why do we make the choices we do and how can we make both the rational and irrational choices better?

PSYCHOLOGY OF CHOICE

If you want to understand the psychology of choice, don't look to architectural journals, planning conferences, or political science researchers; look to marketing, sales, and negotiation. No other fields have so embraced the irrational mind. In the following pages are some key concepts that affect our decisions.

"The Magic Number Seven"

The greater the complexity (and the less time available to unravel the complexity), the more likely it is that the irrational will determine the outcome. This was demonstrated in a 1957 study called "The Magic Number Seven."[4] Research subjects were given a number to remember. They were then asked to walk to another room and repeat the number. Two factors were added to make the task more complex. First, some subjects were given a small number, others a large number that would be harder to remember. Secondly, they were interrupted as they walked from the first room to the second room by one of the research team who thanked the subject for participating in the study. As a reward, subjects were offered the choice between a piece of rich chocolate cake or a bowl of ripe fruit. It was assumed that the rational mind would encourage choice of the healthier option. Those subjects trying diligently to remember their long complex numbers routinely picked the chocolate cake. Those that had the easy-to-remember simple numbers routinely picked the fruit. The rational mind was so distracted by the challenge of remembering the complex number (i.e., the magic seven-digit number), that it made an irrational choice.

Guidance

Keep the options as simple as possible and beware a decision maker who is distracted. Alternatively, if the options are complex by necessity or the decision maker is overwhelmed (as many busy individuals are), use irrational methods to support the option that has (rationally) been predetermined.

The Coin Toss

If you were asked to make a $100 bet on a coin toss, would you take the bet at even odds of 1:1 (win pays $100)? What if the odds were in your favor at 1.25:1 (win pays $125)? Or at 1.5:1 (win pays $150)? Would you risk $100 to win $150?

The rational mind says the risk of a 50/50 coin toss is certainly worth it at *any* odds greater than 1:1. In fact, studies show that some people will never accept the risk ("I'm not a gambling man"), while the average person will only take the bet at 2:1 odds (win pays $200).[5] Why is this? Quite simply, losing is twice as painful as winning is pleasant.

Guidance

Emphasize the downside risks of the worst options more than the upside benefits of the best options. Seek to equalize the risk among the options by eliminating the default (which is always perceived as the least risk) from the options list.

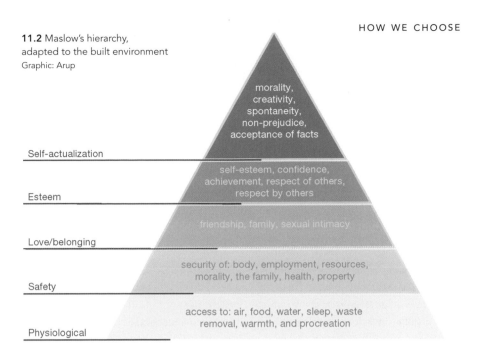

11.2 Maslow's hierarchy, adapted to the built environment
Graphic: Arup

morality,
creativity,
spontaneity,
non-prejudice,
acceptance of facts

Self-actualization

self-esteem, confidence,
achievement, respect of others,
respect by others

Esteem

friendship, family, sexual intimacy

Love/belonging

security of: body, employment, resources,
morality, the family, health, property

Safety

access to: air, food, water, sleep, waste
removal, warmth, and procreation

Physiological

The Value of $50

Would you travel across town to save $50 on a $200 appliance? Absolutely. Would you travel across town to save $50 on a new $20,000 car? Absolutely not. Why not? The value of the $50 is the same. The inconvenience is the same, just a drive across town. And yet the psychology of choice shows that value gets diluted as the price rises. The reason for this actually rests in the way people think about numbers. Although our rational mind knows that 9 is twice as great as 4.5, our irrational mind disagrees. It "knows" that 9 is twice as great as 3! because, it thinks *logarithmically*.[6] So the better question is whether you would travel across town to save $5,000 on a $20,000 car. Absolutely.

Guidance

Don't sweat the small stuff. Position the best choice as relatively small within the context of the greater whole.

So if most decisions are in fact irrational and decision makers often don't disclose (or aren't even aware) of the real reason behind their decisions, how can decisions be made more effectively? Luckily, there's an equation for that.

C = MAT

Decision makers must be motivated to choose effectively, but even those who are highly motivated will make poor choices if their ability to make the good choices is limited (e.g., they have no authority or expertise). Conversely, decision makers with all the ability to make the best choices will nonetheless not do so unless motivated. And lastly, even with adequate motivation and ability, a trigger is still needed to catalyze the motivated action.

Choice therefore equals the product of *Motivation*, *Ability* to act, and a *Trigger* ($C = M \times A \times T$). This formula is adapted from research into behavioral change by Stanford Professor B. J. Fogg, a leading thinker on behavior change theory, founder of the Persuasive Technology Lab, and originator of the Fogg Behavior Model (FBM).[7] We will address Motivation, Ability, and Triggers in the following pages.

Motivation

Are you motivated to do as others do, even more than you're motivated to do what you think you should do? You may say no—you may hope not—but the research actually says emphatically yes… and if you listen closely enough, you will probably hear your inner voice agreeing.

To make effective choices, it's important to realize that what we value publicly is not always what we value privately. People are often unwilling to disclose values that are not culturally accepted or are self-oriented (i.e., self-centered). Motivators are also emotionally biased, based on pain/pleasure, fear/hope, acceptance/rejection.

To understand what we value, whether stated or not, there are good frameworks. The FBM is one. Imagine a design workshop where the question is not "What do you value?" but instead "What are competitors doing or trying to do?" "What would make us happy at the end of this project?" "What could go wrong?" "What gives you pleasure?" Some of these would undoubtedly prompt a laugh, but given that they come closer to core motivations, they are likely to be more effective at answering "What do you value?"

Another way to frame what motivates us was first theorized in 1943 when Abraham Maslow published "A Theory of Human Motivation."[8] The resulting "Maslow hierarchy" presents our needs as a progressive pyramid—basic at the bottom to transcendent at the top—with basic needs requiring fulfillment before higher order needs can be addressed.[9] Figure 11.2 shows an adapted version of the hierarchy of needs as related to the built environment. Maslow's hierarchy has been criticized as ethnocentric (expressing the needs of individualist societies like the United States more than collectivist societies like China or Sweden) and some research has suggested that the hierarchical framing may not have the underpinning that the pyramid ascribes to it. However, Maslow's work still functions as a valuable reminder of the many needs that interweave with our values, as well as the prioritization that often exists between them.

Yet another framework stems from the research by Robert Cialdini, Regents' Professor Emeritus of Psychology and Marketing at Arizona State University.[10] His Principles of Persuasion, adapted to the built environment, are as follows:

- *Reciprocity* People tend to return a favor. Therefore, be a team player.
- *Commitment and consistency* If people commit, orally or in writing, they are more likely to honor that commitment. Therefore, it is important to record project or program goals and ideas early, so that they support decision making later. Even better is storytelling. People who learn stories self-identify with them and repeat and defend them.
- *Social proof* People will do things they see other people doing. Therefore, assuming others are doing good things (like using less energy or improving city health), this can be a positive force in encouraging better choice.
- *Authority* People will tend to obey authority figures, even if they are asked to perform objectionable acts. Don't try to convince everyone to make the best choice, convince the right people to make the best choice.
- *Liking* People are easily persuaded by people they like. Therefore, have fun and make sure others are having fun too.
- *Scarcity* Perceived scarcity will generate competition. The competitive spirit can move decision makers strongly toward good decisions. For example, look at the early adoption of green building rating systems around the world. The initial success was a result of the scarcity (and corresponding novelty) of rated buildings and the prestigiousness of the achievement (later success owed more to the principles of social proof and authority).

Ability

The ability to make good decisions depends on many factors. The guidelines that follow, which can enable better choices, are based on the insights and experience of the authors as well as the FBM.

Simplify

Make the criteria of choice simple enough but no simpler. There's art in understanding how simple a choice should be. We have only one fast rule—there should be fewer than seven criteria. More than the Magic Number Seven and you are treading on irrational ground and must be prepared for irrational tendencies to dominate.

Eliminate the Default Condition

Reduce the risk of making a poor decision by eliminating the default condition, which often doesn't meet the minimum value-based goals. As a result, you can often exclude it from the options set entirely.

Dilute/Diversify

Dilute the risk of a decision within a greater effort (where the decision has already been made). There are often opportunities to combine options that have excellent financial performance (lighting upgrade or weatherstripping) with options that have more questionable financial performance (e.g., renewable generation). By commingling the two, the result can be a deep efficiency improvement with good financial return. The picking of low-hanging fruit has been a missed opportunity, not because it wasn't right to pick, but because so much was still left on the tree.

Timely Integration

Make decisions early (but not too early), and allow enough time (but not too much time). Just as it takes money to make money, it takes time to make time. Integrated design requires greater upfront time commitment, but it saves time, errors, and changes later on. Too often, decision making is poor when the option has been considered for decision too early to be relevant, too late to be integrated, or too quickly to be understood.

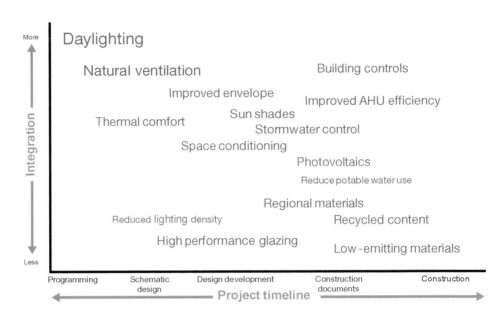

11.3 Items with greater integration impacts should be studied early in a project, while other topics of importance can be detailed later Graphic: Arup/Boora

Invite Skill

Integrated design and planning can challenge those who have to invite the right attendees: if there are too many nothing gets done, but with too few, the missing input can be a lost opportunity. As well as having knowledge in their own field those who are present during an integrated design or planning session need the ability to think outside their immediate expertise. They need to see themselves and their ideas as part of a holistic solution.

Manage Money

Everyone would agree that an adequate budget is needed to pay for the decision. However, it is more important that there is a cost management strategy that provides context for the decision. Managing the *first cost hurdle* is discussed in the next section.

Money, time, and skill are limited, and there are always trade-offs.

Triggers

Fogg describes triggers as either "sparking," "facilitating," or "signaling."

- *Sparking triggers* These help inspire hope or sense of commitment (i.e., support motivation). A beautiful rendering of a finished design or plan is a sparking trigger, as is a financial incentive program or code mandate.
- *Facilitating triggers* These help convince the decision maker that the choice is easy and effortless (i.e., support ability). A slide with a clear recommendation is a facilitating trigger, as is an offer of technical support "to make the process easy."

- *Signaling triggers* These serve as reminders when motivation and ability are already present in sufficient amounts. Think of them as the red/yellow/green traffic lights of design and planning. A follow-up meeting note or reminder of an upcoming review deadline can be good signals.

Numerous triggers occur throughout the design and planning process. "What plan/concept is preferred?" or "Is this site suitable?" are examples of effective triggers.

The range of exploration into more effective triggers is increasing. In December 2009 the New York City Council enacted a series of laws intended to trigger "Greener, Greater [Existing] Buildings" in the city. Each of the laws addressed a different aspect of improving energy efficiency in the city's buildings: energy conservation standards for building renovations; annual energy benchmarking and disclosure; mandatory lighting system upgrades and tenant submetering; and mandatory energy auditing, retrocommissioning, and retrofits. Similar efforts are underway in other cities, states, and countries.

THE FIRST COST HURDLE

Olympic hurdlers often clear a 42-inch (107-centimeter) hurdle by less than an inch, and many times not even that, as evidenced by the hurdles knocked down by feet, knees, and calves. And they do this running as quickly as they can toward a finish line that they often can't see, but know is there.

Clearing the hurdle of affordability can feel just as scary, difficult to predict, and potentially painful. And it all occurs in slow motion.

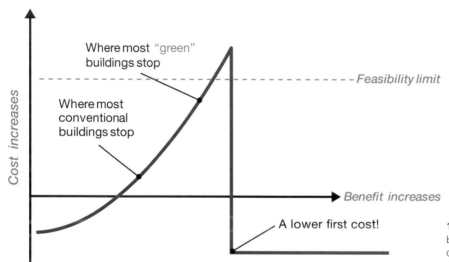

11.4 The concept of "tunneling through the cost barrier" by eliminating costs systemically
Graphic: Arup, adapted from Lovins, Lovins, and Hawken, 2000[i]

When it goes well, though, it seems both effortless and elegant. Decisions are made with reasonable confidence, ambiguity is accepted, risks are managed. In our experience, the projects that have achieved this have had some specific elements in their strategy. We have distilled that strategy into four parts.

Move Money Within the Fixed Budget

Money sloshes around projects until the time when checks are written or contracts signed. And 99 percent of the decisions typically won't have a detailed cost/benefit or life-cycle cost analysis (LCCA). Successful projects therefore rebalance their choice of investments through value-based decision making and the motivational factors discussed previously.

Tunnel Through the Cost Barrier

The beauty of integrated design is the potential to realize cost-saving *synergy*, not just cost *efficiency*. Pushed aggressively, entire systems can be downsized or eliminated because they're no longer needed.

Set a Premium Based on Precedent (e.g., < 2 Percent of Project Cost)

Third-party studies by government,[11] cost estimating firms,[12] and industry associations[13] allow *top-down* budgeting according to performance expectations. For example, an owner can now say that they expect to pay no more than a 2 to 5 percent premium for a LEED Gold building. The result is that an owner, faced with a higher-than-expected estimate at an early stage of design, can maintain their commitment with confidence that their team can converge the cost premium to a reasonable level.

Cost Check Throughout the Process

Once a budget is set, teams can work to achieve the targeted premium. *Bottom-up* cost checking should be used to ensure continued convergence to the targeted budget. An early cost estimate or economic study is invariably high on work involved in resetting the default condition. Actual costs are almost always less. The convergence occurs for several reasons, including: cost estimate clarifications (if estimators don't understand a design, they often put

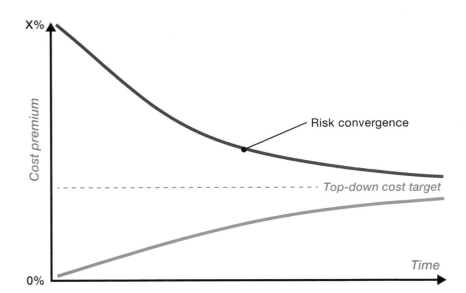

11.5 Cost convergence over time through bottom-up cost checking
Graphic: Arup/Cole Roberts

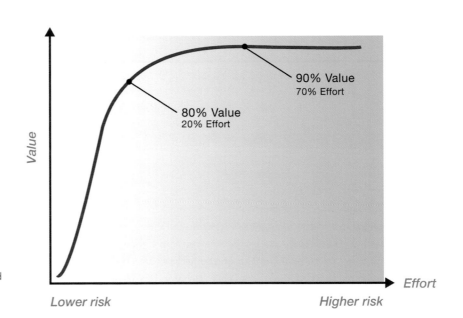

11.6 The 80/20 rule expanded to accommodate risk and value, relative to effort
Graphic: Arup/Cole Roberts

higher numbers to it), teams achieving efficiencies and synergies as their work progresses, and value-based investments (decision makers making good decisions).

EVALUATING OPTIONS AND SETTING THE DIRECTION

The process of evaluating options is more like painting a picture than solving an equation. The best choices emerge out of a diversity of colors and strokes. They feel right, and elegant, but are sometimes unsettling. They recognize that information is informative, not directive; that it is

less important that the data are precisely right, than that the data are adequately approximate and correct. Knowing how approximately correct is correct enough, and when, can be the challenge.

The authors have adapted the 80/20 rule to relate the need for precision and accuracy in proportion to risk (Figure 11.6).[14] The higher you build the farther you can fall, therefore more investment in safety is appropriate. The same relationship is true when evaluating options, since the mere act of evaluation requires effort.

Effective prioritization and appropriate rigor are therefore important in a successful decision-making process. Prioritization allows limited

LCCA Decision Matrix

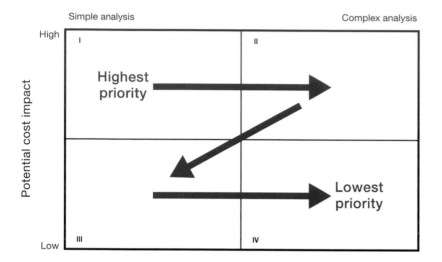

11.7 To prioritize design options for LCCA, a matrix like this can be used. High-cost options that are simple to analyze (I) are prioritized over low-cost options that are complex to analyze (IV)
Graphic: Stanford University Land and Buildings, 2005[ii]

resources to be applied at the most beneficial time and on the most important elements. It avoids unnecessary distractions and helps ensure that the three-legged stool of risk, value, and effort is balanced.

The romance of analysis is its promise of a *right* answer. This romance can compel some decision makers to desire an analysis (e.g., LCCA) for every significant cost that has a potential for future savings. Ironically, those costs *without* the promise of future savings are spared the analytical questioning, especially when they have elements that are aesthetically or culturally prized and hence default. The reality is that the process of analysis is itself a *cost* that reduces the *benefit*. Just carrying out the analysis can tilt the balance of value away from cost effectiveness, so it is best to have the skills needed to undertake a detailed analysis, but use those skills rarely. Even on good projects, less than 1 percent of decisions are analyzed using detailed LCCA or cost-benefit analysis.

A further risk in detailed analysis is the tendency to "paint by numbers," i.e., to make decisions based on the numerical outcome of an analysis. This can seem highly rational, but is often irrational in practice and is used as cover for inadequate decision-making ability. Although the quantitative application of numbers to options can paint a picture, it will often tell only part of the story. The numerical conclusions that arise out of quantitative analysis typically ignore intangible factors and externalities. They also confuse accuracy (correctness) and precision (rightness). As a result, whenever an analysis is produced and reviewed, it should, first, include discussion of intangibles and externalities, and, second, demonstrate its degree of ambiguity. Both of these requirements are discussed in more detail later in this chapter.

There are many methodologies for evaluating options. They attempt with varying degrees of success and suitability to organize, prioritize, and understand the available options. We have collected here four of the most valuable methodologies for making decisions about climate change strategies in the built environment.

Methods

The methods described here are generally arranged in the order that they appear in practice, from early stage to late stage, less detailed to more detailed, many options to few options, and less effort per option to more effort per option.

Criteria-Based Mapping (CBM) ($50–$500 per option)

This is an excellent tool to use at early stages of the design and planning process when, typically, very large numbers of options are brainstormed. Value-based criteria are used to score the various options and plot them on a two-dimensional axis. Additional criteria can be used to color, shape, and size the plotted points. Although CBM can accommodate up to two hundred options, is highly graphical communication, and is a good first filter for prioritization, it has limitations. CBMs are often highly subjective, can be overwhelmed by too many options with similar scoring, and are not an effective final filter for decision making.

BLiP diagrams

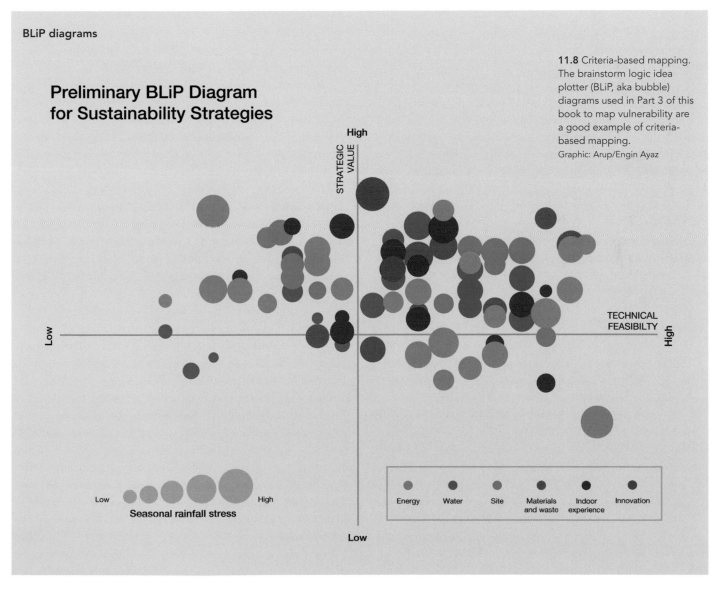

Preliminary BLiP Diagram for Sustainability Strategies

11.8 Criteria-based mapping. The brainstorm logic idea plotter (BLiP, aka bubble) diagrams used in Part 3 of this book to map vulnerability are a good example of criteria-based mapping.
Graphic: Arup/Engin Ayaz

High

STRATEGIC VALUE

TECHNICAL FEASIBILTY

Low

High

Low

Low   High
Seasonal rainfall stress

Energy   Water   Site   Materials and waste   Indoor experience   Innovation

Low

Gap Analysis ($50–$1000 per option)

Sometimes it's more valuable to know what you don't know than to know what you know. A gap analysis is an excellent tool to evaluate an option or organize multiple options. Its goal is to highlight the gaps in knowledge that may signal missed opportunities or significant risks.

Approach Diagrams

The approach diagrams for carbon mitigation, energy mitigation, and adaptation used in this book are examples of gap analyses (see for example Figure 8.3). Often there is an inappropriate amount of effort invested in one step of the approach (e.g., self-generation through renewable energy) and too little effort invested in another step (e.g., load reduction). Each diagram can be used to organize and prioritize dozens, or even hundreds, of strategies. A further benefit is that the approach diagrams also communicate a typical prioritization (e.g., reducing loads before self-generating).

SPeAR

Arup's Sustainable Project Appraisal Routine (SPeAR; see Figure 4.2) is often used as a gap analysis. Each quadrant has lesser wedge subdivisions that are scored based on agreed Key Performance Indicators (KPIs). By evaluating an option through the scoring of each wedge, it is often the case that new, previously unnoticed opportunities are identified. A further benefit of SPeAR is that it can be used to compare multiple alternatives and track changes over time through subsequent updates.

Weighted and Unweighted Matrices ($500–$2,000 per option)

In selecting between a small set of options (fewer than seven), a matrix-based approach can often be put to good use. On one axis of the matrix, the options are listed; on the other, the value-based criteria. Each option is then scored by its value relative to the other options. Relative value is often controlled by holding the total points for all options constant, so that an increase in one value forces an equal decrease to another. The scoring is typically carried out by experts in the topic (e.g., engineers will typically score the efficiency-based performance of renewable energy generation options).

A further level of insight may be added by weighting the value-based criteria, since some values are prioritized over others. This numerical weighting is typically scored by the owner or stakeholder group from which the values originated. The resulting multiplication of professional scoring and value-based weighting can then be summed to a total for each option.

Life-Cycle Cost Analysis ($3,000–$20,000 per option)

LCCA is the only method that can assess the economic return of a given strategy and begin to form the basis for investment grade analysis. It is typically used on a small set of options (two to six) that have reasonably detailed cost and savings estimates. Due to the effort typically involved in developing an LCCA, it is generally reserved for high-cost, high-value decisions where the cost of analysis is equal to or less than the first year potential savings.

Since the quality of the analysis is shaped by the degree of detailed information available, a judgment call is needed to determine how early or late in a design process the analysis should occur. Too early, and the quality of information is inadequate. Too late, and the critical path of decision making is impacted. More than any other method, LCCA can be misused and misinterpreted. We will explore the pitfalls and potential of LCCA in the next section.

The Pitfalls and Potential of Life-Cycle Cost Analysis (and the Dangers of "Payback")

If you present or are presented an LCCA and the answer distills down to one number (e.g., Payback = 5 years, ROI = 32 percent), toss the answer in the trash because it's worthless. Or at least take it with a *big* grain of salt, because the only way to really understand

Whole Building to Net Zero Energy

The six-step net zero energy approach introduced in Chapter 6 and detailed in Chapter 7 was used as a framework for a whole-building NZE LCCA at Stanford University in 2005. The analysis summed the net present value of costs and benefits for building energy measures. The measures were then binned into each of the six steps so that each step would have its own LCCA analysis (Figure 11.9). Finally, all measures were summed as a group and a whole-building LCCA was plotted (the grey line in Figure 11.9). After a year of building occupancy, the actual energy use was examined and the LCCA results re-run with the new data (the red line). The building's desirability had led to very high occupancy and usage that resulted in *greater* marginal energy savings. As a result, the original discounted payback of 4.5 years or 16 percent ROI was improved to approximately 2 years or 30 percent ROI.

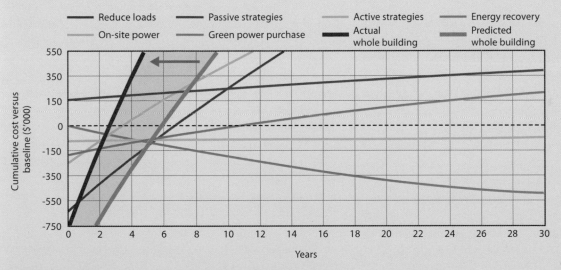

11.9 Stanford Y2E2 building, whole-building LCCA (predicted in gray and actual in red), along with individual performance steps along the six-step net zero energy approach. Discounted payback of all measures (including renewable energy and green power) ranged from two to six years; individual steps had widely varying returns with some not financially strong (e.g., on-site power) and others paying off instantaneously on first-cost basis alone (e.g., passive strategies)
Graphic: Arup/Cole Roberts

an LCCA is to play with the assumptions, test its sensitivity, explore some scenarios, and so on.[15]

Sensitivity, scenario testing, and goal seeking are cornerstones to determining a good answer using an iterative LCCA process. This is because LCCA is often less impacted by what is known today than by the projections and estimates of value decades into the future. Those estimates include (1) capital cost, which due to the often early stage of analysis is conservative and relatively inaccurate, (2) savings, which are only as good as the model accuracy, (3) escalating or deflating commodity expectations, (4) risk judgment and commensurate discount rates, etc.

As a result:

- *Test sensitivity* Knowing how small changes in inputs change the results can be an eye-opener and prompt greater effort to dial in the confidence surrounding assumptions for the most sensitive inputs.
- *Run optimistic, pessimistic, and likely scenarios* If an analysis shows positive returns even under pessimistic assumptions, it can build significant confidence in decision makers.
- *Goal seek* If a particular option is preferred, ask "What would it take to make the decision a 'yes' for this strategy that we think is attractive but that doesn't quite pencil out?"
- *Interpret the graphic results* If a net cash flow line is relatively flat, it has a low ROI and a payback period that is likely to be very sensitive to numerous inputs. If a net cash flow line is relatively steep, it will have a high ROI and be relatively insensitive to inputs.

UNDERSTANDING FINANCIAL IMPLICATIONS AND RISK

"What's your MARR?" What a great first-date question! Right between "What's your family like?" and "What are your hopes and dreams?" More than almost any question, understanding someone's minimum acceptable rate of return (MARR) can tell you a lot about a person, their business, their appetite for risk, their financial optimism, etc. Let me explain through examples. The MARR of the United States federal government for energy-efficiency investments is equal to the published Federal Discount Rate for life-cycle cost analysis, nominally

3.2 percent. Stock market enthusiasts often quote a MARR roughly equivalent to the historic rate of return of the stock market (> about 8 percent). Developers of speculative real-estate developments and products (which include a great many private companies) have an even higher MARR (12–18 percent). Venture capitalists are even more aggressive with MARR expectations—greater than 30 percent—because so few of their investments succeed. The catch is that all too often the MARR that people quote (also often called the discount rate) is not the one they should, because they forget that a MARR is qualified by the risk or uncertainty of the anticipated future cash flows. For a risky investment, the MARR should be high. For a low-risk investment, the MARR should be low.

Fortunately, many investments in climate change mitigation or adaptation are actually low risk because certainty of outcome is relatively high. Energy efficiency returns from energy retrofits are correspondingly certain and low-risk investments when compared to inventing new equipment and products. Scientists' predictions of sea-level rise, storm intensity, and other climate change risks are correspondingly certain and low risk when compared to the world's financial markets and debt-ridden nation states.

This need for certainty is part of the reason that incentive and

What Is a Discount Rate?

The interest rate used in discounted cash flow analysis to determine the present value of future cash flows. The discount rate takes into account the time value of money (the idea that money available now is worth more than the same amount of money available in the future[…]) and the risk or uncertainty of the anticipated future cash flows (which may be less than expected).

Investopedia[16]

Why Should I Care?
If the discount rate is too high, future benefit is devalued and short-term thinking is more likely to prevail.

11.10 The Minimum Acceptable Rate of Return (MARR) is the same as the Return on Investment expectation. High-risk investors (e.g., venture capital) will tend to the left side of the graphic, low-risk investors (e.g., federal and local governments) will tend to the right
Graphic: Arup/Cole Roberts

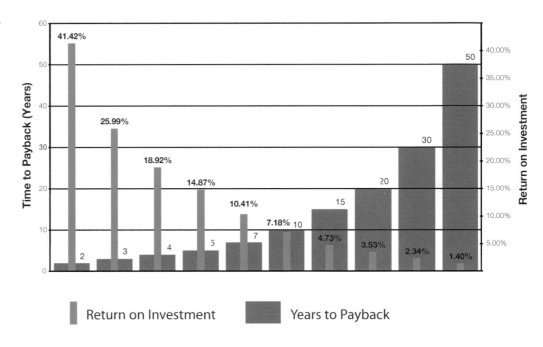

legislative consistency is so important. If an investor was certain that wind energy production incentives were going to be maintained for fifteen years, they would have a different opinion of the risk of the investment (and corresponding MARR) than if those incentives came up for legislative re-authorization every four years. The progressively stringent requirements over long time frames that are contained in the world's first binding greenhouse gas emissions legislation (e.g., Assembly Bill 32 in California) are important not just because they set stringent requirements, but because they forecast the requirements decades into the future.

Intangibles and Externalities
In addition to understanding the risks and return expectations of an investment, it is important to incorporate intangibles (and externalities) into any LCCA results presentation. Indeed, these factors often outweigh the tangible factors of a detailed analysis. Intangibles are typically incorporated into a good decision-making process in

two ways: through (1) agreement on monetary value ranges, or (2) assessment of qualitative value.

Monetizing intangibles may sound counterintuitive, but in our experience a team can often come to a reasonable consensus on a value range. It starts with everyone agreeing that the value is greater than $1 and less than $1 million (for typical intangibles). From there, a facilitating process of converging through discussion to low-end and high-end values occurs. The resulting values can then be included as first cost value offsets or annually recurring value offsets. It is surprising how small the range in value sometimes turns out to be.

The alternative to monetizing intangibles is a reasonably rigorous recording and discussion of them. Each intangible is described in terms of its value to the decision-making group and presented in tandem with the financial LCCA results. The decision makers then judge qualitatively whether the intangible factors are great enough to justify a decision that is contrary to (or supportive of) the LCCA results.

The Dangers of "Payback"

Too many investments proceed with only a cursory review of the simplified payback. Although the majority of these decisions are often well justified (simplified payback can be a good screening tool), a significant percentage will not result in an optimized return for the owner. The reason for this failing is that payback calculations look only for the date at which break even occurs, a full recovery of the investment. They do not consider costs and benefits that may occur preceding or following this break-even point, the rate of return, or the budgetary implications. As the graph demonstrates, the best investment may not be that with the fastest payback.

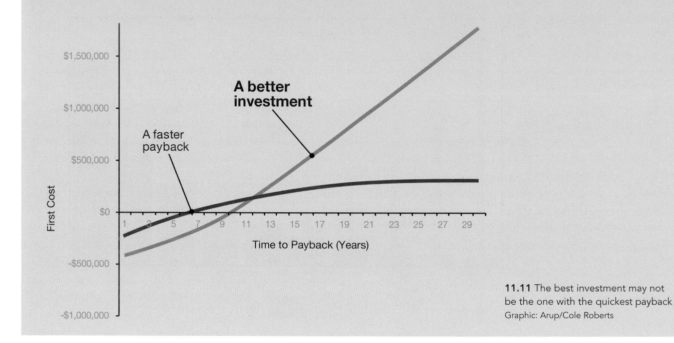

11.11 The best investment may not be the one with the quickest payback
Graphic: Arup/Cole Roberts

Bundling

Understanding risks and intangibles is not enough for good LCCA-based decision making. A third facet, often missed, is the importance of bundling.

Take a moment to study how orchards are harvested. You'll notice that the fruit is picked evenly from all branches according to its ripeness. Over a season, each tree is picked many times. We've handled energy conservation and investments in the exact opposite way, picking only the lowest-hanging fruit and letting the rest of the investment opportunity wilt away. The result is that attractive returns of 5–15 percent (fourteen- to five-year payback) are left on the tree because they weren't quite sweet enough. To leave low-risk investment like that behind today, when safe investments have negative interest rates and the stock market is struggling to make any gains at all, is a major missed opportunity.

The other result is that the low-hanging fruit, which often pays back in under a year and has a financial ROI greater than 40 percent, cannot be used to justify the deeper investments that are needed to respond to climate change. ROIs of 5 percent are left on the tree and the trouble of getting a ladder out to harvest them is not easy to justify. Instead of picking the top options, the best strategy is to bundle multiple options to achieve a target ROI (which should just beat the MARR). There are multiple benefits to this approach. The soft costs are diluted by a larger investment. The net savings are greater (in real currency) since the investment is larger. The risk is reduced since there's a larger diversity of measures. And lastly, deep efficiency gains of 40 to 60 percent often result, in lieu of the old rule of thumb of 10 to 20 percent savings.

Finance

Of course, all of this discussion assumes that there is adequate funding to pay for the measures that are more expensive up front but more financially attractive in the long term. That is why it is important to have a rigorous cost management strategy akin to that described earlier. That is also why it is so important that innovative financing mechanisms like Managed Energy Service Agreements

(MESA), Property Tax Assessed Clean Energy Financing (PACE), On-bill Financing, Power Purchase Agreements, public–private partnerships, and Mortgage Underwriting Energy Risk Adjustments are aggressively established, expanded, and optimized.

TAKING THE LONG VIEW

Deciding to start any effort may be easiest if the first steps are small. Such pilots are simple, diluted, low cost, etc. Establishing a pilot allows flexibility for improvement over time and assumes that optimization will occur. The eco-district systems presented in Chapter 8 are a perfect example, since building such large interconnected and potentially complex systems is impossible to do all at once. The framework of "*establish, expand, optimize, maximize*" allows for more effective decision making *over time*:

1. Establish pilot systems: Build initial phases of future systems that are just big enough to function
2. Expand: Grow the systems over time through good decision making and available investment
3. Optimize: As scale increases, opportunities to tune the system increasingly arise
4. Maximize: Let the system find its optimal scale of operation.

The default condition changes over time. Much of what we treasure today was not built using LCCA, computer-assisted evaluation methodologies, or detailed computational analysis. It was built with an eye toward beauty, quality of construction, suitability, and resilience. Though it was built at the lowest order of Maslow's hierarchy, it achieved some of the highest. We would not go so far as to say all decision making should return to the time when rational argument and detailed analysis were the exception to the irrational exuberance of decision makers; however, a scan through images from our National Historic Register and of World Heritage Sites is a reminder that the correct answer is often different from the right answer.

EVALUATING STRATEGIES AND TRADE-OFFS IN SUMMARY

All great civilizations make decisions. Those that succeed make the right ones. Although a significant opportunity for better decision making occurs simply through integrated design and planning, it's not enough.

Irrational decisions are more common than rational decisions by two to one. And the default condition is the most common "choice" of all. In order to make better decisions, the psychology of choice is important.

Complexity and lack of time can lead to irrational decisions (the Magic Number Seven), so keep it simple. Losing is twice as painful as winning (the Coin Toss), so emphasize the downside risks of the worst options and eliminate the default as an easy option. And position the best choice as small within the greater whole (the Value of $50).

All choices require a combination of *Motivation*, *Ability* to act, and a *Trigger* (C = M × A × T), all occurring at the same time. *Motivation* requires a recognition of values and the emotional bias of individuals. Valuable frameworks include Maslow's Theory of Human Motivation and Cialdini's Principles of Persuasion, each of which is adapted in this chapter to choices made about the built environment. The *Ability* to make good decisions is dependent on simplicity, default availability, dilution of risk, timing, skill, and money management. *Triggers* can spark (motivation), facilitate (ability), or signal (red, yellow, or green light) action.

Since perceived and real costs are among the greatest hurdles to successful decisions, they must be well managed. A four-part management effort includes (1) rebalancing within the fixed budget, (2) tunneling through the cost barrier, (3) setting a premium expectation based on precedent (top down), and (4) regular cost checking (bottom up).

In order to evaluate options successfully, knowing how correct is correct enough, and when, can be a challenge. Prioritize. Align risk, value, and effort. And when detailed analysis is applied, remember that the picture will often tell only part of the story. Intangible factors and externalities are invariably left out. Search out accuracy (correctness) and beware precision (rightness).

Some valuable methodologies for making decisions include criteria-based mapping, gap analysis, weighted matrices, and LCCA. Each has its strength and weakness. LCCA is too often overemphasized. Indeed the only way to really understand an LCCA is to play with the assumptions, test the sensitivity, and explore some scenarios. Is the discount rate (aka MARR) correct for the level of risk? Many investments in climate change mitigation or adaptation are lower risk since the certainty is relatively high.

By bundling strategies in LCCA, soft costs can be diluted, net savings are greater, the investment is diversified, and deep efficiency gains often result. Of course, such an evaluation assumes that there is funding to pay for the measures, so innovative financing mechanisms should be aggressively established, expanded, and optimized, just like significant investments of all kinds.

Looking back on what we have built is perhaps the best guidance for what lasts—quality, suitability, resilience, and beauty.

NOTES

1   Jared Diamond, *Collapse: How Societies Choose to Fail or Succeed*, New York: Viking, 2005.
2   London: Methuen & Co., 1926.
3   "Not in my backyard."
4   George A. Miller, "The Magical Number Seven, Plus or Minus Two: Some Limits on Our Capacity for Processing Information," *The Psychological Review*, 63, 1956: 81–97.
5   Radiolab, WNYC Public Radio, "Choice," 2008. This radio program does an excellent job of demonstrating the wariness of people to taking on risk. Even when that risk is good for them.
6   Radiolab, WNYC Public Radio, "Numbers," 2009.
7   See http://behaviormodel.org/. The website contains numerous valuable resources that are highly recommended for additional reading. The Fogg Behavior Model (FBM) is Behavior Change = MAT. We have used (C)hoice in lieu of (B)ehavior for consistency with the discussion here of how choices are made. The intent is the same.
8   *Psychological Review*, 50(4): 370–96.
9   For a discussion of the details, criticisms, and complementary research that has succeeded Maslow's original work, refer to http://en.wikipedia.org/wiki/Maslow%27s_hierarchy_of_needs (accessed February 26, 2012).
10 For a discussion of Robert Cialdini's work, refer to http://en.wikipedia.org/wiki/Robert_Cialdini (accessed January 12, 2012). Also see http://www.influenceatwork.com/.
11 Steven Winter Associates, "GSA LEED Cost Study: Final Report," Contract No. GS–11P–99–MAD–0565, Norwalk, CT/Washington, DC: Steven Winter Associates, 2004.

12 L. Matthiessen, "Costing Green," report, Davis Langdon Adamson, USA, 2004; "The Cost and Benefit of Achieving Green Buildings," report, Davis Langdon Adamson, Australia, 2007; "The Cost of Green Revisited: Reexamining the Feasibility and Cost Impact of Sustainable Design in the Light of Increased Market Adoption," report, Davis Langdon Adamson, USA, 2007.

13 M. Lucuik, W. Trusty, N. Larsson, and R. Charette, "A Business Case for Green Buildings in Canada," report, Industry Canada, 2005, http://www.cagbc.org/AM/PDF/A%20Business%20Case%20for%20Green%20Bldgs%20in%20Canada_sept_12.pdf (accessed June 25, 2012).

14 The 80/20 rule was first offered up by Joseph M. Juran as a way to benchmark adequate effort. He suggested simply that 80 percent of the value came from 20 percent of the effort (*Architect of Quality: The Autobiography of Dr. Joseph M. Juran*, New York City: McGraw-Hill, 2004). For a more detailed discussion of the 80/20 rule, see the Pareto principle: http://en.wikipedia.org/wiki/Pareto_principle.

15 The exception to the argument to more broadly explore the results of an LCCA occurs when the decision maker has no appetite for the added complexity of the information (e.g., Decision Fatigue, the Magic Number Seven). Sometimes, it is best to just present a single answer as long as the analyst has explored the LCCA behind the scenes so that the single answer presented is a good one.

16 "Discount Rate," n.d., http://www.investopedia.com/terms/d/discountrate.asp (accessed June 20, 2012).

i Paul Hawken, Amory B. Lovins, and L. Hunter Lovins, *Natural Capitalism: The Next Industrial Revolution*, New York: Little, Brown & Company, 2000.

ii "Stanford Guidelines for Life Cycle Cost Analysis," Stanford: Stanford University Land and Buildings, 2005, http://lbre.stanford.edu/sem/sites/all/lbre-shared/files/docs_public/LCCA121405.pdf (accessed June 23, 2012).

Linking growth, job creation, and the economy is key to understanding
climate change investment

Image: fotohunter/Shutterstock.com

12

POSITION CHAPTER

Can We Afford a Low-Carbon Economy?

Simon Roberts

Never let the future disturb you. You will meet it, if you have to, with the same weapons of reason which today arm you against the present.

Marcus Aurelius (121–180 CE), *Meditations*

Against the Earth Summit in 1992, George Bush Sr. forcefully declared, "The American way of life is not negotiable."[1] Almost a decade later, George W. Bush's line objecting to the Kyoto Protocol was "I will not accept a plan that will harm our economy and hurt our workers."[2]

The British economist Professor Nicolas Stern was charged with examining the economic aspects of climate change. His comprehensive review published in 2006 lists historical reductions in national emissions,[3] though most of those given feature a mere 1 percent reduction per year, for example through nuclear power (France), biofuels (Brazil), and the "dash for gas" switch from coal power (United Kingdom). Only one example achieved a greater annual reduction, but this was a consequence of recession: in the former USSR, the economic transition and the associated downturn during the period 1989 to 1998 saw fossil fuel–related emissions fall by an average of 5.2 percent per year.

Given this single historical precedent from the Stern Review of significant emissions reduction, the Bushes have a point. The question in the title to this chapter must acknowledge issues of way of life, the economy, and workers.

This chapter introduces a framework to help us examine these issues. The following structure shows that we can indeed afford a low-carbon economy—one focused on a service sector that can actually create employment and improve quality of life while simultaneously reducing carbon and improving resilience.

A QUESTION OF MONEY

Tackling the question of affordability takes us into the realm of money, but money is just a means of transaction. On the proverbial desert island, a wad of bills would be useful only for fueling a fire to cook over or bunching up as insulation to keep warm. (Way more useful would be a solar electric panel complete with a set of electric tools.)

On the other hand, no one can deny the power money bestows when received as a salary: to buy goods and services, settle bills, pay the mortgage, etc. Or in the business world, consider a discussion between a specialist sustainability designer and their client developer planning a new building. The designer would list means of energy efficiency, improved occupant comfort, and microgeneration by on-site renewables. But the client would say these all have to stay within a cost ceiling of 10 percent over current market standards, otherwise the building would not be affordable and the developer would go out of business. Money matters.

So what kind of framework considers whether society as a whole can afford the transition to low carbon? It should examine two aspects: whether the economy has the capacity to make the required investments and how to prevent such policies from being ruinous to the economy.

The framework introduced in this chapter gets to the heart of the capacity for an economy to make the mitigation changes described in this book. For each investment proposition, it helps distinguish and navigate several aspects:

- Economic affordability
- Futile or extravagant investments
- At what point the investment is simply down to society's will and desire to make the commitments
- Behavior changes for reduced emissions that have minimal systemic consequences.

The framework considers a national economy as a whole and takes into account its systemic nature. Dependencies between the various parts mean that the consequences of changes in one ripple through to others, manifesting in surprising ways. Such a framework can form the basis of a system dynamics computer model[4] configured to test investment propositions and run what-if scenarios.

BUILDING A SYSTEM VIEW

The starting point for this framework is a system perspective involving such terms as *growth*, *gross domestic product* (GDP), *consumption*, and *investment*. These are probably familiar and might have negative associations. After all, is it the need for incessant economic growth that holds us back from climate change mitigation measures? Is it individuals' wanton demand for ever more material consumption?

GDP has received bad press. How can it be that so many—politicians, business leaders, in fact most members of society—subscribe to the need for economic growth? I invite you to set to one side any prejudice about GDP, or the thinking "This is economics—I won't understand it!" GDP is an internationally agreed system of national accounts having great rigor and uniformity, developed over many years, to provide accurate insight into the fundamental behavior and performance of economies. Historically, wars and threats of war have provided the main impetus for the development of national accounts.[5] This method of accounting was seen as a quantitative framework for devising policies to mobilize a nation's resources to fight wars or to repair the subsequent damage. What if we viewed addressing climate change as a similar challenge?

THE OUTPUT FORM OF GDP

Let's begin by looking at two facets of GDP data: output and expenditure. Output is what companies and other economic units of an economy do—the production part of GDP. The technical phrase is to *add value* or, in its full form, *gross value added* (GVA). It is the dollar value for the amount of goods and services that have been produced, less the cost of all inputs and raw materials that are directly attributable to that production. When GVA is divided by number of staff employed, this is the productivity of employees.

Figure 12.1 introduces a schematic for the output concept. It shows three larger sectors of the economy—manufacturing, construction, and services—by boxes that represent their physical assets. The assets are the buildings, the equipment they contain, and the vehicles used (all known as *capital stocks*). This type of schematic is a Sankey diagram, in which the width of the colored band emerging from each box represents the volume of its output, its GVA. The example is for a large industrialized nation, but the approach is globally applicable to many economies.

For businesses to add value, they each need a range of inputs, such as imports, direct fuels, electricity, and jobs. These are shown

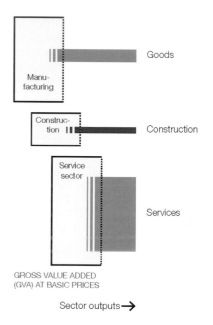

GROSS VALUE ADDED
(GVA) AT BASIC PRICES

Sector outputs ➜

12.1 The output volumes of GDP for the three larger sectors of an example economy
Graphic: Simon Roberts, Data: UK Office for National Statistics (ONS), 2010

in Figure 12.2 as if flowing from the left into each of the boxes and metamorphosing into the sector's output.

In this example, the volume of imports for the manufacturing sector is much greater than that made domestically, substantially swelling the goods band that continues to the right. This sector is also a substantial user of petroleum products, electricity, and gas, but it employs comparatively few. The construction sector has no imports and is a low energy user, but it employs almost as many as manufacturing does. The service sector imports only a small proportion and is a comparatively low consumer of electricity and gas given its large volume output (GVA), but it is clearly the dominant employer. The service sector covers distribution, hotels, transportation, communications, finance, business, public administration, defense, education, health, and social work.

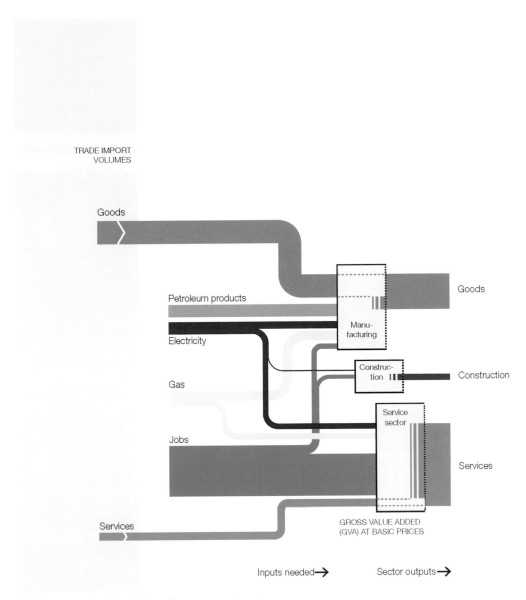

TRADE IMPORT
VOLUMES

Goods

Petroleum products

Electricity

Gas

Jobs

Services

Manu-
facturing

Construc-
tion

Service
sector

GROSS VALUE ADDED
(GVA) AT BASIC PRICES

Goods

Construction

Services

Inputs needed→   Sector outputs→

12.2 Inputs required by the three larger economic sectors
Graphic: Simon Roberts, Data: ONS, 2010 [ii]

THE EXPENDITURE FORM OF GDP

Expenditure is at the far end of the whole production process within a country. It consists of final supply, fulfilling final demand, which is finally bought—but what counts as the "end"? The formalism of GDP is clear on this point. There are four types of final destination:

- Direct purchases by consumers
- Government services for the benefit of consumers paid for indirectly (by taxation)
- Exports
- Investments.

National accounts show how final demand is supplied by goods, services, etc., as illustrated in Figure 12.3 (overleaf). (This schematic also introduces flows from the smaller economic sectors of agriculture and utilities.) Note in this example that most construction ends up going to investment and services contribute to all end uses.

An important point to make about the expenditure end is that it is demand-led, meaning it determines the flows that enter the system from the left. For domestic consumption, consumers choose to buy—though the level of "choice" might be arguable. Households buy essentials such as food and luxuries such as a foreign holiday, but advertising has a huge influence on today's luxuries becoming

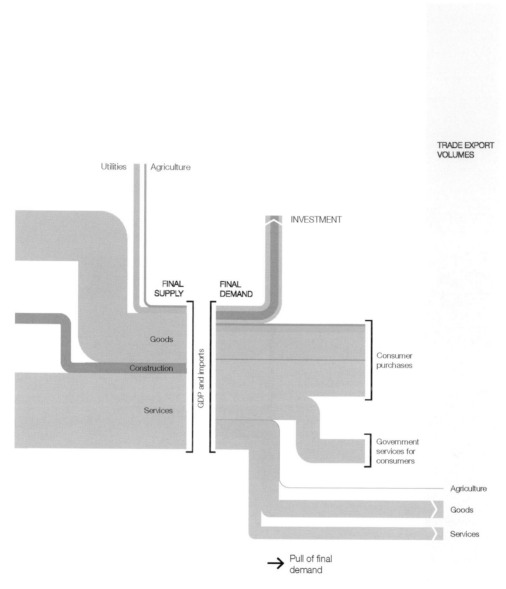

12.3 The expenditure form of GDP, showing final supply of agriculture, utilities, goods, construction, and services, divided between four types of final demand
Graphic: Simon Roberts, Data: ONS, 2010[iii]

"indispensable" tomorrow. Of the 12.6 percent of Americans below the federal poverty line, 80 percent have an air-conditioner, 75 percent have a car or truck, and 33 percent have a computer, dishwasher, or second car.[6] Nevertheless, "the market is king," with consumers making the final purchasing decisions.

The other three components of final expenditure are also demand-led. Government services used by consumers are ultimately determined by elected representatives through democratic accountability (getting voted back into office). Exports depend on other countries choosing to buy the goods and services offered.

Investments respond to what needs to be built—more on that later.

THE FULL SYSTEM SCHEMATIC

The output and expenditure forms of GDP introduced above will now be placed within the system schematic, with final consumption on the right and various inputs, such as jobs and fossil fuels, on the left. Three more aspects are needed to complete the system schematic:

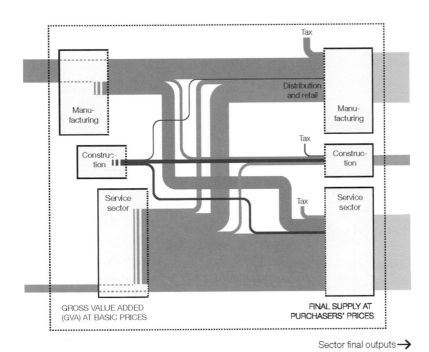

Tax

Distribution
and retail

Manu-
facturing

Manu-
facturing

Tax

Construc-
tion

Construc-
tion

Service
sector

Tax

Service
sector

GROSS VALUE ADDED
(GVA) AT BASIC PRICES

FINAL SUPPLY AT
PURCHASERS' PRICES

Sector final outputs →

12.4 Intermediate or intrasector demand on the right is supplied from the left;
also on the right is tax on products, as paid by purchasers
Graphic: Simon Roberts, Data: ONS, 2010 [iv]

- Intermediate demand
- Other physical assets
- Investment in physical assets (capital stocks).

In this context, domestic production is considered output and termed *GVA*. Expenditure at the end is defined as *final demand*. In between is *intermediate demand*, which covers transactions taking place between all parts of the economy. For instance, manufacturing provides furniture and electronic equipment to the service sector to enable it to function. The service sector provides payroll and advertising services to manufacturing. Figure 12.4 shows this intermediate demand between the principal economic sectors. The schematic actually shows each principal sector as a pair of boxes on the left and right, each box having a dotted border for the side facing the other. For simplicity, the schematic is arranged as if something is produced on the left and input from another sector is added later. (This is not actually the case since the complexity of the economy is about many interactions between businesses at all stages through production.) While GVA is normally given at basic, producers', or factory-gate prices, expenditure is given at final or purchasers' prices, which include taxes on products. (Taxes appear in the schematic as an input because their effect is to increase the cost.) Since final supply on the right is substantially different from each sector's GVA on the left, the bands change color as they emerge on the right.

In addition to physical assets of the three larger sectors, other physical assets to include are as follows:

- The smaller economic sectors of agriculture and extraction
- Oil refineries that convert crude oil into petroleum products, such as gasoline, diesel, jet fuel, lubricants, and naphthas

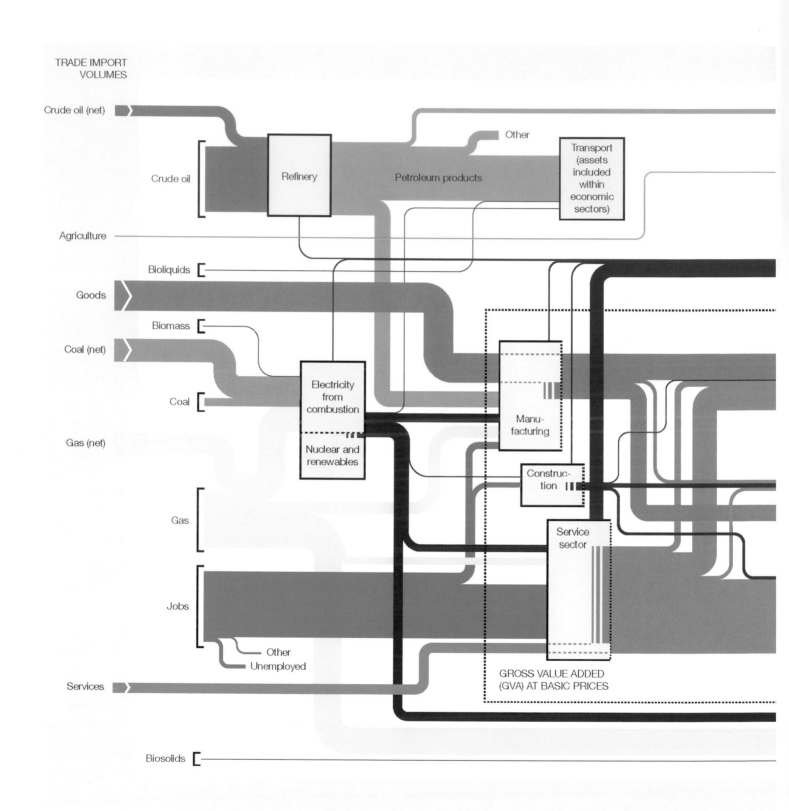

TRADE IMPORT
VOLUMES

Crude oil (net)

Crude oil

Refinery

Other

Transport
(assets
included
within
economic
sectors)

Petroleum products

Agriculture

Bioliquids

Goods

Biomass

Coal (net)

Coal

Electricity
from
combustion

Manu-
facturing

Gas (net)

Nuclear and
renewables

Construc-
tion

Service
sector

Gas

Jobs

Other
Unemployed

GROSS VALUE ADDED
(GVA) AT BASIC PRICES

Services

Biosolids

12.5 Full system schematic showing flows of energy, jobs, goods, construction,
and services (yellow boxes represent physical assets, the rectangular limit is the
physical extent of the economy and trading border)
Graphic: Simon Roberts, Data: ONS, 2010[v]

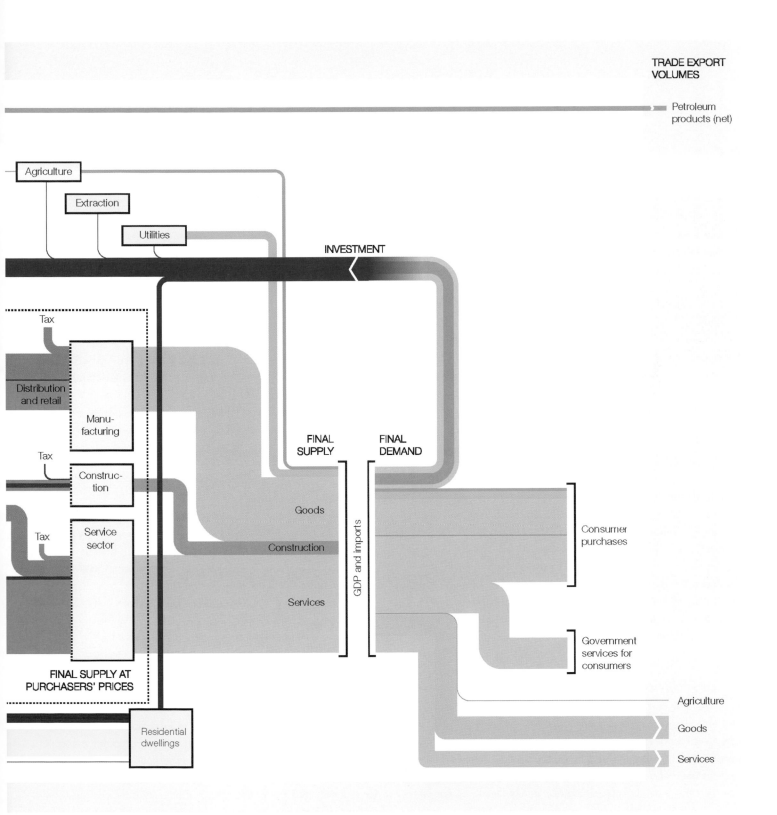

TRADE EXPORT
VOLUMES

Petroleum
products (net)

Agriculture

Extraction

Utilities

INVESTMENT

Tax

Distribution
and retail

Manu-
facturing

Tax

Construc-
tion

Tax

Service
sector

FINAL
SUPPLY

FINAL
DEMAND

Goods

Construction

GDP and imports

Services

Consumer
purchases

Government
services for
consumers

FINAL SUPPLY AT
PURCHASERS' PRICES

Residential
dwellings

Agriculture

Goods

Services

- Power stations that convert fossil fuels and nuclear fuel into electricity, together with renewable energy from hydro, solar, and wind
- Transportation, a direct user of fuels and electricity
- Residential dwellings.

All physical assets depreciate over time, so they require investment. Assets that must expand to meet increasing demand need additional investment. Earlier investment was framed as a component of final expenditure (Figure 12.3). It is this same investment that feeds each physical asset represented by the boxes across the schematic.

Figure 12.5 is the fully assembled system schematic. It is enclosed by a rectangular border defining the system boundary of the physical extent of the country in which the physical assets reside, crossed by the trading of fuels, goods, and services that make up imports and exports. The investment flow can now be seen doubling back from final demand and connecting to each of the physical assets. (The size of GDP is shown at the point in the system where its sum with imports equals final demand.)

In Chapter 2, Figure 2.2 shows the inventory of U.S. greenhouse gas emissions and sinks from 1990 to 2009 for five parts of the economy. The system schematic here shows how these five parts can be traced back to just the source fossil fuels responsible for their emissions ("Industry" and "Commercial" in Figure 2.2 equate to "Manufacturing," "Construction," and "Services" here).

EXAMPLE PATHS THROUGH THE SYSTEM

As an example, let's start at the bottom left with someone employed in the service sector. They work in a building that consumes gas and electricity, and needs investment to maintain its fabric. If we suppose the building is a retail outlet, we would then follow the services band across and up as it merges with goods. The goods now have the retail markup or margin, and might finally end up with consumers on the right.

In times of recession when fewer goods are purchased, we can follow back along the goods path for the consequences. On the one hand, there are fewer imports—good for the balance of payments.

On the other, reduced employment for those working in the retail sector could be substantial.

The schematic example demonstrates that the service sector is a significant part of the economy:

- Providing the most jobs
- Creating the largest volume of output
- Receiving the most investment.

Economists tend to refer to this as the world economy becoming increasingly "lightweight." One commentator, patently ignoring the input dependencies shown in the system schematic, put it that "as much as two-thirds of global economic activity consists of outputs that don't pollute or even weigh anything at all—things such as entertainment, education, finance, and health care"![7]

HISTORICAL TRENDS

The system schematic is helpful for comparing sizes of flow, but these data are from only one point in time. Over time, volume flows vary—Figure 12.6 (a) and (b) shows that both the number of jobs in the service sector and the output of the sector (GVA) have increased (from 1990 to 2008).

Let's consider how many jobs are needed for each unit of output and the way this number has varied over time. The number is found by calculating a ratio of the two graphs, as in Figure 12.6 (c), which clearly shows how the number of jobs per unit output has been falling continuously over the entire period. The line derived from historical data is almost indistinguishable from a smooth trend line.

This falling ratio makes evident the continuous pressure for any organization in the service sector to reduce the number of jobs. This is in order to reduce costs and thus stay competitive and in business. Luckily, the size of the service sector as a whole has been growing to such an extent that this growth not only compensates for reducing jobs per unit output but actually increases the total number of jobs.

How has the sector managed to grow? The full system schematic reminds us that demand originates on the right side. As long as demand rises, then the increase propagates back upstream. One consequence

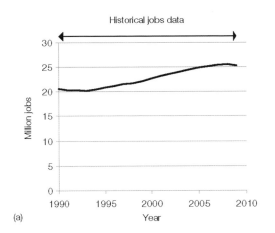

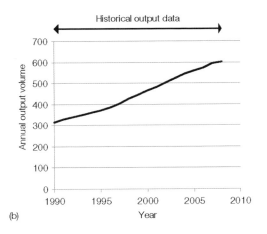

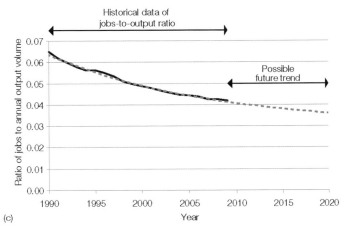

12.6 Historical data for the service sector of (a) jobs and (b) annual volume output units; (c) is the ratio of jobs to annual output volume with a best-fit trend curve extended forward to 2020
Graphic: Simon Roberts, Data: ONS, 2010[vi]

is that unemployment is kept down, which is essential—but substantial investment is needed, with the service sector receiving the widest investment line in the diagram.

FEATURES OF THE FRAMEWORK

The full system schematic (Figure 12.5) links four concepts:

- All forms of energy, on the left
- Physical assets, which fulfill all the roles of the economy
- Jobs
- Final demand of consumers and exports.

The framework promised at the start of this chapter consists of this

systemic view, which is extended by time variations of relationships from historical data and future trends, as shown in Figure 12.6.

Economists often reduce the complexity of the economy down to one metric of *growth*. The schematic shows that too much is lost in that approach. We need to see the component flows, their interactions, and consequences. Others who focus on climate change policies also have a simplified approach of shrinking the economy down to one metric of *embodied carbon*, as shown in Figure 12.7. This is also a very limited analysis. For instance, a consequence of consumers simply reducing purchases of high carbon content might be reduced economic activity, leading to higher unemployment. We need a better approach.

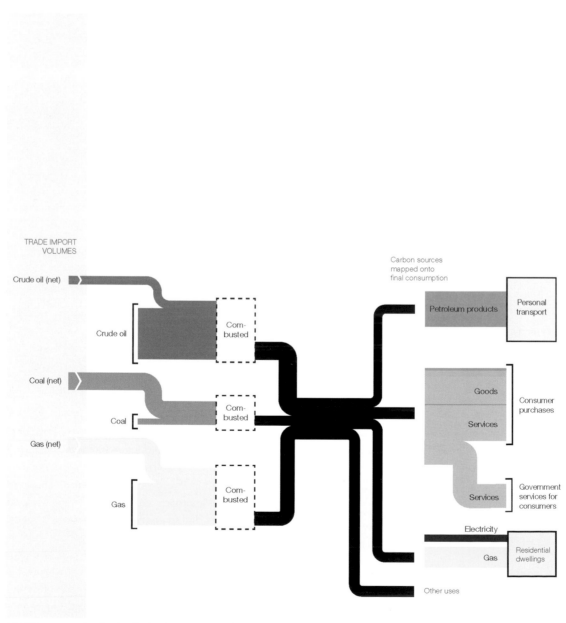

TRADE IMPORT VOLUMES

Crude oil (net)

Crude oil — Com-busted

Coal (net)

Coal — Com-busted

Gas (net)

Gas — Com-busted

Carbon sources mapped onto final consumption

Petroleum products — Personal transport

Goods — Consumer purchases

Services

Services — Government services for consumers

Electricity

Gas — Residential dwellings

Other uses

12.7 The approach of embodied carbon consists of mapping carbon sources onto final consumption and other uses (including personal transportation and residential dwellings)
Graphic: Simon Roberts, Data: ONS, 2010 [vii]

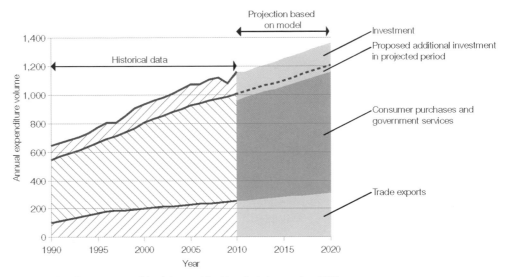

12.8 Components of final demand for historical data and to 2020.
The extrapolation includes a proposal for slightly reduced consumer
consumption to enable greater investment, as would be necessary for
measures needed to shift to low emissions
Graphic: Simon Roberts, Data: ONS, 2010 [viii]

HOW CAN WE AFFORD A LOW-CARBON ECONOMY?

This section briefly describes how the framework can be applied to
exploring emissions reduction options in the economy and society.
First, consider the following proposed hierarchy of compliance for
an economy:

1. Jobs are essential
2. Jobs are mainly provided by an expanding service sector
3. Therefore demand for services must be allowed to grow
4. Therefore a growing service sector must have its investment
   needs fulfilled.

Figure 12.8 shows the historical evolution of final demand
components of GDP. Historical levels of investment must be maintained
to comply with point 4 in the hierarchy. Measures needed to shift
to low emissions must be implemented faster than the market is
currently managing, so they require additional investment. It is plain
from Figure 12.8 that a greater proportion of GDP must be diverted
from consumer consumption to investment than has happened

historically. This is shown on the right-hand side for a possible trend
into the future. (Note how consumer consumption still increases in
the long term.)

The compliance hierarchy can be extended:

5. The manufacturing sector has declining employment
6. Renewable energy reduces dependency on imports for fossil
   fuels
7. Commercial transportation plays a key role in economic activity
8. Private transportation has the least systemic interaction with
   economic activity.

In selecting investment measures for low emissions, it will be essential
to include improved energy efficiency of both the service sector
and transportation.

Within the framework, each investment measure for reducing
emissions requires just three pieces of information:

• Scale and plausible rate of implementation

- Cost of implementation
- Resulting change in emissions (by generation of energy or reduction of demand).

For example, consider investment in building wind turbines. There will be various feasible rates of scaling up turbine capacity, ultimately constrained by the wind resource and land on which to place turbines. Each build rate has an annual investment cost. The turbines generate electricity, which reduces need for fossil fuel.

When the framework is run as a system dynamics computer model, sets of measures can be compared to find those needing minimum additional investment. Runs of the model also show consequential changes to inputs upstream as the mix of investment and consumer consumption varies. Once there is a coherent overall plan with optimal investment, then we move on to the detail of implementation. This can take several forms:

- Government investment funded through taxation or bonds
- Regulatory requirements on industry to invest, which lead to higher prices
- Encouragement of consumer investment, by individuals or communities

Note that from the perspective of the framework, these are all equivalent in diverting final demand to the investment path.

AFFORDING A LOW-CARBON ECONOMY IN SUMMARY

Carbon content is a first step in sensitizing all of us to the impact of behaviors and economic activity. However, alone it doesn't take us forward. It is too blunt an instrument and can't provide a tuned means of developing the transition.

There is a hierarchy of demands upon the economy, the highest being employment. The nature of our economies—free market, OECD (Organisation for Economic Co-operation and Development) economies—creates an incessant pressure to minimize costs that leads straight through to reducing the need for jobs. The service sector must continue to expand. This should not be presumed to represent "bad consumption." (The quality of those jobs and work is a different agenda, not covered here.) The economic solution to maintaining overall employment is that new services are created that consumers choose (or can be persuaded) to purchase.

In principle, we have the knowledge to decarbonize electricity, erect buildings that demand zero energy, and remove any need for fossil fuels through clever technology. However, the investment demands would be just too high, with widening of the investment portion in Figure 12.8 gobbling up too much of the consumer consumption band.

Targeted investments can be identified, however. The framework introduced here provides the means for developing a low-carbon transition that is affordable.

NOTES

1 Jack Beatty, "Playing Politics with the Planet," *Atlantic Unbound*, April 14, 1999, http://www.theatlantic.com/past/docs/unbound/polipro/pp9904. htm (accessed June 10, 2012).

2 CNN, "Bush Firm Over Kyoto Stance," *CNN.com*, March 29, 2001, http://edition.cnn.com/2001/US/03/29/schroeder.bush/index.html?iref=allsearch (accessed June 10, 2012).

3 *Stern Review: The Economics of Climate Change*, 2006, http://webarchive. nationalarchives.gov.uk/+/http:/www.hm-treasury.gov.uk/independent_ reviews/stern_review_economics_climate_change/stern_review_report. cfm (accessed June 8, 2012).

4 Simon Roberts, "Arup's 4see Model: Combining Socio-economic and Energy Aspects of a Whole Economy," *Research Review* 3: 20–3, http://www. arup.com/~/media/Files/PDF/Publications/Research_and_whitepapers/ Research_Review_250112lr2.ashx (accessed June 10, 2012).

5 François Lequiller and Derek Blades, *Understanding National Accounts*, Paris: OECD Publishing, 2006, http://www.oecd.org/dataoecd/37/12/38451313. pdf (accessed June 10, 2012).

6 Kate Pickett and Richard Wilkinson, *The Spirit Level: Why Greater Equality Makes Societies Stronger*, New York: Bloomsbury Press, 2010.

7 Charles Kenny, "Everything You Know about Peak Oil Is Wrong," *Bloomberg Businessweek*, January 26, 2012, http://www.businessweek.com/magazine/ everything-you-know-about-peak-oil-is-wrong-01262012.html (accessed June 10, 2012).

i "Input-Output Supply and Use Tables, 2010 Edition," Office for National Statistics, 2010, http://www.ons.gov.uk/ons/rel/input-output/input-output-supply-and-use-tables/2010-edition/index.html (accessed July 2, 2012).

ii Ibid.
See also "JOBS02: Workforce jobs by industry," Office for National Statistics, 2012 http://www.ons.gov.uk/ons/rel/lms/labour-market-statistics/june-2012/index-of-data-tables.html (accessed July 2, 2012).
See also "Annual tables: 'Digest of UK energy statistics' (DUKES)," Department of Energy & Climate Change (DECC), http://www.decc. gov.uk/en/content/cms/statistics/energy_stats/source/total/total.aspx (accessed July 2, 2012).

iii See note i above.

iv See note i above.

v See note ii above.

vi See note ii above.

vii See note ii above.

viii See note i above.

Leadership isn't different; it's the beginning of a new normal
Image: Arup

13

Corporate Leadership

Alisdair McGregor

> In our view, the climate change challenge will create more economic opportunities than risks for the U.S. economy.
> U.S. Climate Action Partnership[1]

THE ROLE OF CORPORATIONS

It is sometimes difficult for elected leaders and governments to initiate, test, or mandate low-carbon technologies. At the national scale in the United States it has proven very difficult to pass climate mitigation policy due to the many well-funded special interests. European countries have had more success in introducing mitigation policy, but when attempts are made for agreement at the global level it is hard to get truly significant policy agreed. At the other end of the spectrum individuals operate at too small a scale for their individual leadership to make a difference quickly, but they can put pressure on corporations. In the absence of policy, corporations (and large institutions) can and should take a leadership position. They are able to do this because they have different motivations and drivers that allow them to provide leadership in technical solutions, attitudes, and communication. They also have relatively tight management controls that allow them to measure and monitor their performance. Of course corporations can move in the wrong direction and act as a force for derailing climate mitigation strategies, as already discussed in the first part of this book. How can we demonstrate that moving to a low-carbon business model is the best strategy for the environment and financial security?

Large corporations are financially very powerful and cross national boundaries. Indeed the larger corporations have economies larger than many countries. During the recession of 2008–11, many corporations increased their profitability and cash reserves. It was reported that Apple had more cash reserves than the U.S. government in 2011. The leadership of corporations is not elected, which gives them more flexibility and nimbleness for rapid change. Although shareholders have an influence on the leadership it is a very different control from that of an electorate. Recent SEC (U.S. Securities and Exchange Commission) disclosure guidance and shareholder claims are proving that investors need more information with regard to climate change exposure. CEOs can have enormous influence not only on their own companies but on the markets in which they operate. As demonstrated in Chapter 14, direction from Walmart CEO, Lee Scott, was essential in launching Walmart on the path of reaching a sustainable business operation. A similar approach was taken by Jeffrey Immelt at General Electric (see box overleaf).

Corporations touch everything we do. Almost everything we buy is produced or delivered by a corporation. Corporations have the ability to communicate at the point of purchase, but the consumer also has the same ability. This leads to a balance of influences. How much should corporations take the lead in offering more sustainable products to a consumer that may be interested only in getting the lowest price? Walmart started placing compact florescent light (CFL) bulbs in the prime, middle shelf location. Their goal was to increase the penetration of these energy saving bulbs into a wider market. On the other hand increased consumer activism and purchasing choices are causing corporations to move toward more sustainable products. In the early days of the U.S. Green Building Council's LEED program there were few material choices that met the requirements of LEED. Now there is a vast choice of sustainable materials available.

There are reasons beyond cost why people make buying or renting decisions, and many of these can be linked to intangibles. In the commercial real-estate market productivity and employee attraction are exceptionally large drivers, along with brand image. These intangibles are hard to quantify but important. Linking these intangibles with financial return on real-estate developments is covered by Scott Muldavin in his book *Value Beyond Cost Savings*.[2] Demonstrating the value of these intangibles to the building owners and developers is a vital role of the design and construction team.[3]

Corporations also interact with local, state, and national governments. There are two very different interactions from the corporate world with elected governance. On the one hand there are corporations who are trying to remove or emasculate environmental regulations and climate policy, and on the other there are those that are moving ahead of current regulations and who want to see stronger regulations to level the playing field. Many believe that increased environmental regulation will deter business from locating in a region or country if they can locate to somewhere with little regulation. Reality would appear to disprove this argument. The San Francisco Bay Area has some of the tightest energy and environment policies in the United States and yet some of the most successful

General Electric Case Study

For many years General Electric (GE) was a target for environmentalists due to its support of coal and nuclear power and a well-publicized case of dumping chemicals in the Hudson River. In early 2006 GE's CEO Jeffrey Immelt launched the company in a new direction.[4]

The goals set in 2006 were to cut overall greenhouse gas (GHG) emissions in 2012 to 1 percent below their 2004 levels. While that might seem a soft target, in 2006 GE had a projected growth rate that would have resulted in GHG emissions at 40 percent above 2004 levels by 2012. Jeffery Immelt promoted the change with the tag line "green is green," convinced that investing in the development and promotion of clean and green technologies would be profitable. The company also publicly proclaimed climate change to be real and stressed the importance of acting now. Prior to the launch of the new strategy, GE spent eighteen months getting feedback from customers. The message they received was that demand for cleaner technologies was driven by the following:

- Rising fuel costs
- Tightening environmental regulations
- Consumer demand for green products.

Were these just bold words for a marketing campaign, or would GE deliver real change? GE's Citizenship Report for 2010 reports on a number of goals and policies that cover both social and environmental sustainability.[5] Progress on GHG reduction is shown in Table 13.1.

So, entering 2011, GE had exceeded its initial target of a 1 percent reduction in GHG emissions by 2012 and was already at 24 percent below its target. Some of the overall reduction may be due to the global economic slowdown. A more meaningful target was the reduction in energy intensity (the amount of GHG per unit of economic activity). In 2006 the target was to achieve a 30 percent reduction on the 2004 baseline by 2008. At the end of 2010 they had reached a 33 percent reduction, but are targeting a 50 percent reduction by 2015.

The financial success of GE's strategy is less clear. Measured against the average performance of the Dow Jones Industrial index, GE has not performed that well, although there are many other issues that have affected GE's stock performance.

TABLE 13.1
GE's progress on greenhouse gas reduction
Graphic: Arup

2010 commitments	Progress	2011 commitments
Continue long-term GHG and energy-use reduction trend and drive to the following goals: • 50% improvement in energy intensity by 2015 (2004 baseline) • 25% reduction in GHG emissions by 2015 (2004 baseline)	GE continued to make progress on these goals; GHGs were reduced by 24% and energy intensity improved by 33% from the 2004 baselines. During 2010, GE hosted its first-ever Forum on Industrial Energy Efficiency, and devised a strategy to drive GHG and energy-use reduction in GE's business units.	Engage business leaders in driving GHG and energy intensity use reductions. Implement an ecomagination scorecard for GE's internal environmental footprint against which activities that drive to the goals will be measured.

TABLE 13.2
The spectrum of corporations' responses to climate change
Graphic: Arup

Aggressive Status Quo	Compliance	Sustainable Organization
Maintain status quo at all costs	Comply with environmental regulations	Sustainability is an essential business model
Fund lobbyists to remove environmental regulations	Realize that some response to sustainable issues is required as a license to operate	Green credentials and strategies give a competitive advantage
Environmentalism is harmful to economy	Sustainability is a necessary cost but not a business model	Triple bottom line reporting of operations
Fund global warming deniers		Short- and long-term perspective
Short-term perspective		View not being sustainable as a risk
See sustainability as a risk in itself		

corporations in the world are located there, including Google and Apple. These and similar companies are already moving to a more sustainable operation and their workforces, consisting of young creatives, are demanding social and environmental responsibility.

THE SPECTRUM OF CORPORATIONS

Corporations span the full spectrum in terms of response to climate change and moving to a low-carbon economy. We have summarized the spectrum into three broad groups, as shown in Table 13.2.

We believe that the biggest impact can be made by moving companies in the Compliance zone into the Sustainable Organization zone.

Only regulation and market demand will move the companies in the Aggressive Status Quo zone to the Compliance zone. The science of climate change will not influence this group. If the market demand moves to the sustainable end of the spectrum then they will either go extinct or be forced to change to survive. Examples from this group include Exxon and the Charles Koch Foundation, funding global warming deniers. Texas oil businesses heavily funded a repeal vote of the landmark California carbon reduction legislation, Assembly Bill 32. Fortunately the California electorate voted against the repeal.

It can be argued that all companies strive for some part of the triple bottom line. However, some purely focus on short-term financial returns and engage in greenwash.

To move the Compliance group forward there is a need to demonstrate the business case for going to a low-carbon and sustainable operation. It is necessary to quantify the risk of climate change and show the benefits of sustainable operations as they relate to the specific corporation. Characteristics of Sustainable Organizations include:

- Linking economic prosperity with social and environmental goals
- Willing to lengthen the time horizon used to judge their performance
- Viewing social and environmental issues as potential risks to their prosperity.

The quantification of risk due to not moving to a more a sustainable business model is often overlooked. The insurance industry is realizing the potential liability and insurance losses associated with climate change. Swiss Re was one of the major sponsors of the publication *Climate Change Futures: Health, Ecological and Economic*

Dimensions.[6] Swiss Re has studied the effects of climate change on a number of vulnerable places in the world and arrived at the conclusion that it is cheaper to adapt to a warming world now than to wait and hope. The year 2011 was the worst on record for global economic losses due to extreme weather-related events.

There are companies that are forging ahead and moving the market (Walmart, Toyota, GE, etc.) but they want the playing field leveled; in other words tighter regulations to bring up the laggards. On the other hand, forward-looking organizations respond to incentives better than minimum standard regulations.

The good news is that the number of companies in the Sustainable Organization category is increasing at a rapid pace. In 2012 Haanaes et al. reported on a survey of nearly 3,000 executives from the commercial sector, asking how they are implementing and developing sustainable business practices.[7] Two-thirds of the executives stated that sustainability was necessary to be competitive in today's marketplace. This is an increase from 55 percent in 2010. An increasing number of business organizations are promoting climate change policy. An example is the United States Climate Action Partnership (USCAP), which issued "A Call for Action" in 2012.[8] This paper promotes carbon trading and other incentives to encourage low-carbon industry and calls for regulations to be introduced to penalize emissions. Although the short- and medium-term carbon dioxide reduction targets are lower than recommended by scientists it is encouraging to see major corporations part of the group, including U.S. auto manufacturers and some of the big oil companies.

AVOIDING GREENWASH

It is promising that many major corporations are setting and achieving GHG reductions well ahead of Kyoto targets and national goals. However, there is justified concern that some corporations are seeing this as an opportunity to improve their image and gain market share without making real long-term changes to their operation. The Global Reporting Initiative (GRI) was developed to provide a uniform reporting format and standard for corporations to report their progress on sustainable issues.[9] Companies that don't use GRI can also publish their annual sustainability or citizen reports alongside the usual financial report. Arup now reports on their sustainability performance as well as financial performance in their annual report.[10] In addition, several standards and methodologies have been developed to quantify carbon emissions or footprints for corporations and supply chains. In the United Kingdom PAS 2050 has been developed as the world's first product carbon footprint "standard."[11] It aims to make carbon footprints of goods and services ("products") comparable so that businesses can better realize efficiencies in their supply chains (and resultant cost savings) and consumers making procurement decisions can be more informed about the associated carbon emissions of their purchases.

PAS 2050 is relevant to producers of goods and providers of services, as well as major procurers of products in the private and public sector. It provides a common basis for demonstrating ongoing carbon reduction initiatives, whether they result from the Carbon Reduction Commitment (CRC), Greenhouse Gas Protocol, and/or other corporate carbon reduction aims. Arup has assisted the Carbon Trust with the development of Footprint Expert, a toolkit aimed at delivering PAS 2050 assessments.[12] Footprint Expert has been used to calculate the carbon footprints of over 5,500 stock-keeping units (SKUs) with a combined value of £2.7 billion in product turnover.

CORPORATE LEADERSHIP IN SUMMARY

Corporate organizations have often been portrayed as the enemy of sustainability and carbon reduction policy. What we are seeing is that the majority of corporations are moving towards a more sustainable business model. A significant proportion of those companies that are adopting sustainability see it as an essential component of their business operations. It is somewhat of a mystery why companies that reside in the Aggressive Status Quo group, who are now in the minority, appear to get so much more political traction. As sustainable organizations become the norm for business reasons we may well see that business drives down our carbon footprint faster than government regulation.

NOTES

1   "A Call for Action," USCAP, n.d., 3, http://us-cap.org/USCAPCallForAction. pdf (accessed June 25, 2012).

2   Scott Muldavin, *Value Beyond Cost Savings: How to Underwrite Sustainable Properties*, San Rafael, CA: Green Building Finance Consortium, 2010; see http://www.greenbuildingfc.com.

3   T. Gascoigne, "How Sustainable Real Estate and Green Practices give Businesses a Competitive Advantage," *Smart Business*, December 1, 2011, http://www.sbnonline.com/2011/12/how-sustainable-real-estate-and-green-practices-give-businesses-a-competitive-advantage/?full=1 (accessed June 11, 2012); D. L. Pogue, C. C. Tu, and H. M. Bernstein, "Do Green Buildings Make Dollars and Sense?" presentation given at Greenbuild conference 2010, http://marketing.cbre.com/Sustainability/ GreenBuildingStudy/DoGreenBuildingsMakeDollarsAndSense.pdf (accessed June 11, 2012).

4   "A Lean, Clean Electric Machine," *The Economist*, December 3, 2005, http://www.economist.com/node/5278338 (accessed June 10, 2012).

5   http://files.gecitizenship.com.s3.amazonaws.com/wp-content/ uploads/2011/07/ge_2010_citizenship_report.pdf (accessed June 10, 2012).

6   Paul R. Epstein and Evan Mills, *Climate Change Futures: Health, Ecological and Economic Dimensions*, Boston, MA: Center for Health and the Global Environment at Harvard Medical School, 2005.

7   K. Haanaes, M. Reeves, I. von Streng Velken, M. Audretsch, D. Kiron, and N. Kruschwitz, "Sustainability Nears a Tipping Point: Full Report," *MIT Sloan Management Review* 53(3), 2012: 1–17; see https://pubservice. com/MSStore/ProductDetails.aspx?ID=78879.

8   See note 1 above.

9   See https://www.globalreporting.org/information/about-gri/.

10 See for example http://www.arup.com/Publications/Corporate_Report_2010. aspx.

11 British Standards Institute, "PAS 2050:2011: Specification for the Assessment of the Life Cycle Greenhouse Gas Emissions of Goods and Services," London: BSI, 2011.

12 See http://www.carbontrust.com/client-services/footprinting/measurement/ carbon-footprint-software.

Walmart's team successfully lowered the installed cost of radiant tubing from over $8 per square foot for conventional installations to just $2.50 per square foot by using a prefabricated piping mat that could be rolled out and fixed down just before the concrete slab was poured

Image: Arup/Amit Khanna

14

The Walmart Story

Alisdair McGregor

Walmart opted to answer the [Carbon Disclosure Project] questions for the first time in 2006.... Last year, more than 500 of Walmart's suppliers measured and reported their greenhouse gas emissions to the CDP for the first time.

Keith Littlejohns, The Carbon Disclosure Project[1] (a nongovernmental organization), 2012

In previous chapters there have been many excellent examples of low-carbon and net zero energy buildings. However, in order to make the overall carbon reductions demanded by mitigation requirements low-carbon design needs to reach the mass market. It would be hard to think of a better organization to represent the mass market than Walmart.

Walmart operates over 8,500 retail stores worldwide so even small improvements in energy consumption in each facility have a big impact at the global level. Walmart has been looking at energy efficiency since the 1990s but it wasn't until 2003 that the program really took off. Changing the direction of a large organization takes time and a lot of effort, and requires enthusiasts and activists in the ranks combined with board-level support and direction.

In the mid-2000s, Walmart CEO Lee Scott gave that direction to the company and set it on a course to reaching a long-term sustainable business model.

What would it take for Walmart to be at our best, all the time? What if we used our size and resources to make this country and this earth an even better place for all of us? What would that mean? Could we do it? Is this consistent with our business model? What if the very things that many people criticize us for—our size and reach—became a trusted friend and ally to all, just as it did in Katrina?[2]

Walmart set up a series of sustainable value networks to drive their business toward the ultimate goal of a sustainable business (see Figure 14.1). The complete approach of Walmart to sustainability is a book in itself, but their progress toward their ultimate goal can be reviewed in their 2011 Global Responsibility Report.[3]

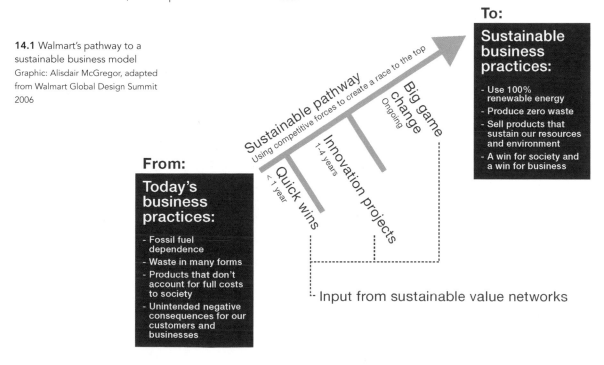

14.1 Walmart's pathway to a sustainable business model
Graphic: Alisdair McGregor, adapted from Walmart Global Design Summit 2006

This chapter will focus on the sustainable facilities value network. The approach to sustainable design in the mass market has many lessons that can be transferred to other building types. The lessons learned will be helpful to other organizations with large building portfolios looking to make major improvements to their environmental footprint.

THE EXPERIMENTAL STORES

Primary Goals

Walmart initiated the experimental stores program in the spring of 2003. The primary goals of the project were as follows:

- Reduce the amount of energy and natural resources required to operate and maintain the stores during a three-year time period following the grand opening
- Reduce the amount of raw materials needed to construct the facility
- Substitute, when appropriate, the amount of renewable materials used to construct and maintain the facility.

Two Supercenters were selected for the experimental program: Aurora, Colorado (dry summer, cold winter), and McKinney, Texas (hot and humid). The schedule for McKinney was tight, so the technologies adopted had to be capable of being incorporated quickly into a 195 Prototype (a 195,000-square-foot store). Aurora, with a longer lead-in period, could incorporate more radical systems and technologies.

The performance of these stores was monitored by two U.S. Department of Energy (DOE) national laboratories: the National Renewable Energy Laboratory (NREL) monitored the site in Aurora and Oak Ridge National Laboratory (ORNL) monitored the site in McKinney.

The design, construction, and monitoring of these projects was a team effort involving many companies. When I received a call in 2003 from Walmart, I had no idea what a long and innovative trip Arup was embarking on.

The Design Process

At the start of the project a series of workshops were held to explore all the possible strategies and systems that might be applicable to Walmart stores. Ideas were then grouped into site and building categories. The ideas were ranked according to how well they matched the primary goals.

The ideas that passed this initial filter were then entered into a master spreadsheet, which became a working document to further filter ideas. Costs, potential payback, and notes were added as information became available. The spreadsheets were reviewed with Walmart and then formed the basis for design of the two superstores. Further refinements took place as the design progressed and better information on cost and return on investment became available.

Large retail stores with a significant grocery component have a different energy use profile from commercial or residential building types. A Walmart Supercenter uses about a third of its floor area for grocery, and the remainder for dry goods. The majority are operated 24/7. A typical energy-use profile for a store is shown in Figure 14.2.

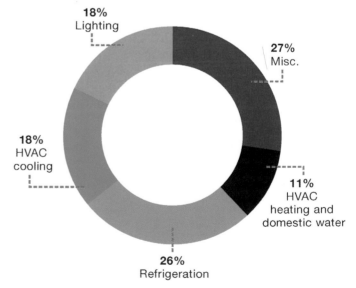

14.2 Distribution of annual energy consumption for a typical Walmart Supercenter
Graphic: Arup

14.3 The solar wall at Aurora
Image: Costea

A great number of systems and materials were tested, both within the two stores and in the parking and landscape areas around them. The next section focuses on the major ideas, both those that did not work as expected and the success stories.

Summary of Store Experiments
Each strategy is annotated with the relevant step from the six-step energy approach described in Chapter 6 (see Figure 6.1).

Refrigeration Energy Saving and Recovery
Refrigeration energy accounts for 26 percent of the annual energy consumption of a typical Walmart store (Figure 14.2). Strategies were adopted to reduce the refrigeration load and to recover the energy for useful purposes.

- *Alternative freezer/cooler refrigeration units (Steps 1 and 3)* Typically only freezer cabinets have doors and refrigerated display cases are open shelf, allowing the refrigerated air to spill out. Glass doors were added to the display cases to reduce the load on the refrigeration systems. All display cases utilize ultra-high-efficiency ECM (electronically commutated motors) fan motors, reducing the total display case energy consumption. Additionally, at the Aurora store, the medium-temperature refrigeration systems have been redesigned to use secondary refrigerant technology with evaporative condensing, which reduces the primary refrigerant (R-404A, a greenhouse gas) usage by 50 percent. The secondary refrigerant, 35 percent propylene glycol (antifreeze), is not under high pressure like the primary refrigerant, so ABS piping is used, reducing the total lineal feet of copper pipe installed by 40 percent in a typical store.

- *Captured waste heat from refrigeration (Step 4)* Excess heat generated by the building's refrigeration system is used to heat water for hand washing and to heat floors in specific areas of the store, offsetting gas demand. This was a strategy already deployed in many stores.

- *Waste to energy (Step 4)* Two sources of waste were investigated as heat sources: waste cooking oil and waste engine oil from the Tire & Lube shop. The oils were collected and stored in separate tanks and then used as fuel sources for a special boiler designed to fire on waste oil. At the time of the design Walmart was paying to have waste oil taken off-site. An initial life cycle cost showed a positive result with off-haul costs removed and gas consumption reduced. Once the buildings opened, the biofuels market had taken off and Walmart could sell the recovered oil. The boilers turned out to be temperamental in operation and frequently broke down. They have now been removed from the stores. This is an example of a "seemed like a good idea at time" strategy.

HVAC Systems
Typical Walmart stores use package roof-mounted air conditioning units that blow conditioned air into the space at high level. The team investigated several options for displacement air supply systems.

- *Solar wall, Aurora (Step 2)* The south-facing wall of the store has a perforated metal siding that draws the air through and heats it with solar energy. The solar-heated air is sent to some of the air-handling units (AHUs) on the roof. The wall raises the temperature of the incoming air by 10 to 20°F (6 to 11°C) so saving the use of natural gas. In the summer when no heating is required the hot air is vented at the top of the wall (Figure 14.3).

- *Passive cooling at the Garden Center (Step 2)* The Garden Center shade structure at McKinney is oriented to take advantage of natural breezes on site. In addition, a cupola with thermostatically controlled louvers and fans was added. As the temperature rises the thermostat opens the louvers, allowing hot air out of the space, while drawing in cooler air near the customers (Figure 14.4).

- *Air distribution system (Step 3)* The sales areas use displacement ventilation, which focuses the cooling and heating system on conditioning only the first eleven feet of the space. Fabric ducts (Duct Sox) are mounted eleven feet above the floor and supply air at low velocity and moderate temperature (typically 65 to 68 °F [18 to 20°C]) through perforations in the fabric. The supply air mixes with the surrounding air and slowly falls to the floor level. Heat gains from people and equipment then lift the air to roof level by convection (Figure 14.5).

- *Radiant floor heating (Step 3)* A radiant floor heating system was installed in specific areas to improve comfort. Hot water pumped through a series of plastic pipes in the concrete floor result in heat radiating from the floor. The radiant heating system transfers warmth that helps shoppers feel comfortable in the cool refrigerated section or the Garden Center in the winter. Radiant heating is also installed in the access pit of the Tire & Lube Express so auto service associates will be comfortable in winter.

- *Evaporative cooling, Aurora (Step 3)* The AHUs that supply the sales floor use indirect evaporative cooling, taking advantage of the dry climate. For much of the year this system can provide the 65°F (18°C) air required for the displacement ventilation system. Additional cooling from the absorption chiller is only required for the hottest times of the year.

- *Cogeneration plant, Aurora (Step 4)* The team investigated several types of fuel cells but the composition of the available natural gas would have caused premature degradation of the cell pack. Instead a packaged cogeneration system was installed consisting of an absorption chiller plant, six 60-kilowatt microturbines (for a combined rated electrical output of 360 kilowatts), and a cooling tower. The system delivers electricity, cooling, and heating to the building and the microturbines run on natural gas. During a utility power outage the microturbines provide back-up power to the refrigeration system.

Lighting Systems

Lighting energy accounts for 18 percent of typical store consumption. Walmart was already using skylights and dimming to reduce energy levels so several options were tested for further improvements.

- *Reduced nighttime lighting levels (Step 1)* At night the light levels in the store are reduced to decrease the light level variation from the outside. This is done for two reasons: (1) to help customers adapt as they come and go, and (2) to reduce energy consumption during the nighttime hours.

- *Focused lighting (Step 1)* By focusing the light on the produce it is more prominent in appearance and the lighting consumes less energy.

- *Active daylighting, Aurora (Step 2)* The receiving area of the

14.6 Inside refrigerated displays, showing the LED lights
Image: Costea

Aurora store uses the Soluminaire. This product is similar to a skylight in appearance but uses a mirrored reflector that tracks the Sun throughout the day. The mirror directs more daylight into the building than a traditional skylight of equal size.

- *Natural daylight and dimming controls (Step 2)* Skylights and clerestories allow daylight directly into the store (Figure 14.5). When sufficient daylight is available, the store lights are dimmed or turned off, reducing energy loads.

- *Main store area lighting (Step 3)* The smaller diameter single T5HO lamps in the store produce as much light as two of the T8 lamps used in previous stores. The color of the lamps, referred to as *color temperature*, has been increased to 5000 K, to match the daylight entering the store. The color temperature contains more blue light than is traditionally used in retail stores, improving the ability to see merchandise in the store.

- *LED lights in grocery cases (Step 3)* LED lighting is used inside the grocery cases with doors, in place of fluorescent strip fixtures. LED lights have a longer lifespan than fluorescent lights, produce less heat, and use significantly less energy than typical grocery case lighting. In addition, florescent lighting is typically seen with a yellow or blue hue. Utilizing 5000 K LED lighting, the true colors of products are viewed inside the case (Figure 14.6).

14.7 Clerestory photovoltaics at McKinney
Image: Costea

On-Site Renewable Energy (Step 5)
Both stores tested various types of photovoltaic (PV) panels with different levels of integration.

McKinney

- *Building-integrated PV: roof-mounted polycrystalline* Polycrystalline PV laminates mounted in the Garden Center's canopy reduce the store's demand on the local electrical power grid (Figure 14.4).
- *Building-integrated PV: roof-mounted, clerestory amorphous* Thin-film PV laminates are integrated into the roof of the entry vestibules.
- *Building-integrated PV: curtain wall-mounted and clerestory-mounted polycrystalline and amorphous* Thin-film and crystalline PV laminates are integrated into the south-facing, front entry facade of the store. Thin-film laminates have been used in the clerestory to allow natural daylight to enter the store at the checkout registers (Figure 14.7).
- *PV: flat roof-mounted thin film* Thin-film PV laminates have been adhered directly to the new roof membrane of the Tire & Lube Express.

Aurora

The Aurora store used the south-facing slope of the north light clerestories to test three types of PV panel.

- *Roof-mounted single crystalline PVs (PowerLight)* These glass/plastic PV laminates are installed as part of a PowerGuard PV system manufactured by PowerLight of Berkeley, California.
- *Roof-mounted crystalline PVs (Rwe Schott-Solar)* Edge Film: Growth PV modules have been installed on the east and west ends of the southernmost roof monitor. These glass/glass PV modules have been installed with a rack mount system attached directly to the upturned seams of the roof monitor's standing metal seam roof. No roof penetrations are used to attach the system to the roof.
- *Roof-mounted amorphous PVs (Uni-Solar)* These UV stable polymer-encased PV laminates have been field applied directly to the flat portion of the standing metal seam roof panel with an integral adhesive. No roof penetrations are used to attach the system to the roof.
- *50-kilowatt wind turbines* 50-kilowatt wind turbines (Bergey) were installed at both stores. The turbine was designed for low average wind speed areas, where wind power had not been practical in the past. Unfortunately, due to a variety of issues including problems with the inverter, the turbines rarely met their predicted level of performance. They did, however, provide a very visible manifestation of the stores' sustainable ethos.

14.8 Photovoltaics and the solar wall: renewable energy for both heating and electricity
Image: Costea

Site and Landscape Experiments

There were several experiments that focused on improved handling and treatment of stormwater runoff, harvesting of stormwater for irrigation, and low water landscaping.

Lessons Learned from the Experimental Stores

Both stores were monitored for performance over a three-year period. The full results from the monitoring would take many pages to cover, but some key points were learned from the data collected that were applied to future stores and can also be applied to all large-scale retail buildings. In both locations a similar-sized Walmart Store with the same geography was monitored to provide a performance comparison.

At both locations the energy use of the comparison stores gradually reduced over the monitoring period. The probable reason is that as the store managers were getting high-quality performance data they were adjusting the operation of the store to reduce energy use. It clearly shows that getting good feedback can reduce energy use without any physical changes in building plant. The overall carbon dioxide (CO_2) output for the experimental stores was lower than the comparison stores, but not as low as hoped for (see Figure 14.9 overleaf).

Some of the reasons for the shortfall in performance have been discussed above. Other reasons were as follows:

- The displacement ventilation was set up as a constant volume system. In a standard store the rooftop air-conditioning units turn off when space temperature is satisfied. This reduces energy use even though there may be less than ideal air movement in the space. The displacement system has since been retrofitted to a variable air flow control, allowing the air supply to be reduced by half under low load. At half flow the Duct Sox still remain inflated but significant energy reductions are achieved.

- At both stores there were difficulties in commissioning and getting the correct control sequences working. At McKinney the performance of the air-conditioning improved in the third year of operation as the systems were brought closer to their design operation. The clear lesson to be learnt for all buildings that use atypical systems is that time needs to be allowed for monitoring performance and adjusting controls to reach design expectations.

- The bank of six microturbines at Aurora did not produce the CO_2 reduction expected. In part this was due to the complexity of the heat recovery and absorption chiller performance, such that the use of waste heat was not maximized. The system has since been decommissioned. A key lesson for retail is to keep it simple.

- The combination indirect evaporative AHUs with pre-heat from the solar wall proved difficult to commission correctly. Once the controls and other factors were corrected the units have performed well, and the solar wall is very effective for a climate with cold but sunny winters.

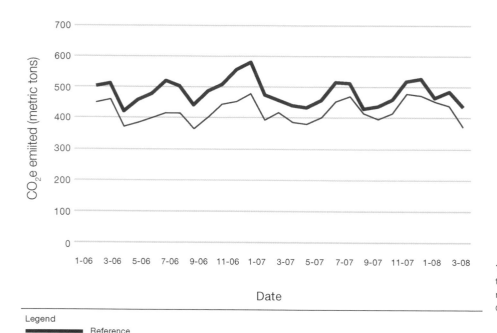

14.9 CO_2 equivalent emissions for the McKinney store, compared to a reference store

Graphic: adapted from NREL report

- In general the PV installations operated close to their predicted performance. As the cost of PV panels is coming down this is clearly a technology well-suited to the large roof areas of big-box retail stores, as well as for shading parking areas.

THE HIGH-EFFICIENCY STORE PROGRAM

Walmart made a strategic business commitment in 2005 to have its stores move toward the corporate environmental goals of being supplied by 100 percent renewable energy, to create zero waste, and to sell products that sustain resources and the environment.

Goal/Objectives

Walmart initiated efforts to increase the efficiency of its stores with the development of the High-Efficiency (HE) design concept. The purpose of the HE store concept is to move toward creating a prototype that is 25 to 30 percent more energy-efficient than the prototype currently built today, and thereby to contribute to overall corporate environmental goals.

This was a very different approach from the experimental stores, which were, as the name implies, a test-bed for a large number of systems. The HE program was set up as an evolutionary process, each new design series learning from the previous and achieving progressively lower energy consumption. Some of the knowledge developed through the experimental store program was transferred to the HE series for further development. The move to hydronic systems is an example of this evolution.

There are some common features between all five of the HE series. The first is to combine the refrigeration and space heating/cooling systems, which in traditional retail design are separate. This requires a water circuit to move energy from one system to the other. In HE.1, 2, and 4, the traditional air-cooled packaged air-conditioning unit is replaced with a water-source heat pump. The refrigeration and air-conditioning system share a common cooling tower for final heat rejection. When heat is needed on the sales floor the heat pumps first use the rejected heat from the refrigeration units and only if there is excess heat in both systems are the cooling towers activated. The system principle is shown in Figure 14.10.

Evolutionary Design

The design of the integrated pump, cooling tower and boiler unit for the HE series is a good example of evolutionary design and shows the importance of continuously striving to simplify and improve the overall performance of a system. The Aurora store had a separate ground-level plant room. For the HE stores we wanted to develop a prefabricated, skid-mounted plant unit that would sit on the roof. The pump package is a central component in the HE design solution. The pump package is a roof-mounted, semi-custom piece of equipment that provides a weatherized and conditioned space for the building condenser water loop components, supplemental heating water boiler, electrical power distribution, and controls.

14.10 Integration of refrigeration
and space conditioning using a
water circuit
Image: Arup

Summer integration

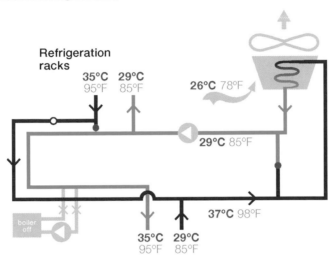

Refrigeration
racks

35°C 95°F   **29°C** 85°F   **26°C** 78°F

29°C 85°F

37°C 98°F

boiler
off

35°C 95°F   **29°C** 85°F

Water source heat pumps (cooling mode)

Winter integration

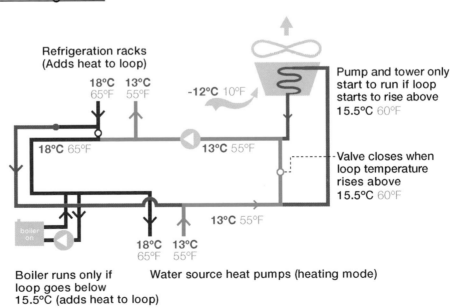

Refrigeration racks
(Adds heat to loop)

18°C 65°F   **13°C** 55°F   **-12°C** 10°F

Pump and tower only
start to run if loop
starts to rise above
15.5°C 60°F

18°C 65°F   **13°C** 55°F

Valve closes when
loop temperature
rises above
15.5°C 60°F

boiler
on

18°C 65°F   **13°C** 55°F   **13°C** 55°F

Boiler runs only if
loop goes below
15.5°C (adds heat to loop)

Water source heat pumps (heating mode)

Legend

Cooling tower on roof

Ambient air

○ Open

Pump

● Closed

HE.1

The HE.1 pump package footprint was approximately 450 square feet and housed four condenser water pumps (fully redundant refrigeration loop pumps and dual HVAC loop pumps), dual cooling tower pumps, one heating water pump, two heat exchangers, a heating water boiler, space for water treatment equipment, interconnecting water piping, electrical power distribution, and building management controls.

HE.2

The progression from HE.1 to HE.2 produced significant changes in the overall design, and specifically within the pump package. The major design revisions within the pump package included:

- Replacing the open loop cooling towers with closed loop fluid coolers, eliminating the cooling tower pumps and heat exchangers, and eliminating a portion of the water treatment system
- Elimination of the fully redundant refrigeration pump; with piping and control modifications this allowed the use of the second HVAC loop pump to back up the single refrigeration pump
- Reducing the amount of interconnecting piping within the package
- Electrical power and control modifications.

The HE.2 design reduced the footprint of the pump package to approximately 170 square feet, nearly 70 percent smaller than HE.1. The smaller pump package yielded first cost savings and the overall design modifications produced energy savings when compared with HE.1. The HE.2 package weighed approximately 55 percent less than the HE.1 pump package, and allowed for cost reduction in the building structure supporting the weight of the package.

HE.4

Continued engineering refinement during the design of HE.4 led to the elimination of the boiler and circulating pump from this prototype design for all climate zones, which produced additional changes to the pump package to further reduce cost and improve energy efficiency. These changes included:

- Reducing the quantity of heat pump units reduced the flow requirement of the HVAC condenser loop. This flow reduction allowed for one of the two HVAC pumps to become a true standby pump for either the HVAC or refrigeration loops. The flow reduction also allowed for a downsizing of the piping within the pump package.
- The heating water boiler and heating water pump were eliminated, which allowed for a reduction in the overall package size.

HE.5 and HE.6: Hot/Dry Climate-Specific Store Design

The HE mechanical systems are intended to positively interact with the prevailing climate and consume the lowest possible amount of energy, while assuring comfortable, safe, and maintainable service for the future. As a broad application, the concept of radiant cooling is used in conjunction with independent fresh air ventilation, with evaporative cooling. Routine low wet-bulb outdoor conditions allow the use of both free air-side and water-side cooling. Unlike a typical retail store that needs up to forty roof top units (RTUs) to heat and cool the building, the HE.5 and 6 use only ten AHUs, which bring in fresh outdoor air to maintain air quality. The reduction in rooftop units considerably reduces noise, raw materials, and maintenance costs (Figure 14.11).

The following key concepts were investigated in the systems' design:

Evaporative Air Cooling

The Southwest U.S. climate is generally hot with routinely low humidity. In other words, the wet-bulb depression (the difference between dry-bulb and wet-bulb temperatures) is mostly high, thus encouraging the use of evaporative cooling techniques. As elaborated below, evaporative cooling has good potential for reducing the cooling load on the chiller.

Indirect Evaporative Cooling (IDEC)

Indirect evaporative cooling is provided by supplying cold water directly from the fluid cooler to the cooling coil in the IDEC unit. Given a wet-bulb approach of 7°F (4°C), the fluid coolers in drier conditions can provide water temperatures in the 50 to 60°F (10 to

14.11 A rooftop clear of air-handling units
Image: Arup

16°C) range. Preliminary analysis shows that indirect evaporative cooling will perform for over 3,000 hours in a year.

Radiant Cooling

Space conditioning is provided by pumping evaporatively cooled water through pipes cast into the floor slab. Keeping the floor temperature in the range of 66 to 68°F (19 to 20°C) will produce comfort conditions even with an outside temperature above 105°F (41°C). There are three stages of cooling:

1. Evaporative cooling only, where radiant floor, AHU circuits, and refrigeration condensers are all fed from the fluid coolers
2. Evaporative cooling in the radiant floor, chilled water supplied to the AHU circuit
3. Chilled water cooling in both the radiant floor and HVAC circuit, refrigeration cooling directly from evaporative coolers.

The Radiant Floor Systems consume approximately 50 percent less electricity annually, compared with the Baseline All Air System,

ALISDAIR MCGREGOR

Using Walmart's Purchasing Power to Drive Innovation

Initial life-cycle cost analysis by Arup for the radiant cooled floor slab showed that the payback period was too long. However, as the energy-saving potential was large, Walmart pressed the piping manufacturer to see if the installed cost of the radiant piping could be brought down from $8 per square foot for conventional installations to below $3 per square foot. The manufacturer came back with the concept of a prefabricated piping mat that could be rolled out and fixed down just before the concrete slab was poured. The labor costs were dramatically reduced, resulting in an installed cost of $2.50 per square foot.

14.12 Low-cost radiant floor installation
Image: Arup

by using higher temperature chilled water and thermal mass benefits to maximize the amount of IDEC. The Baseline All Air System, on the other hand, requires the chiller to provide low temperature chilled water (at about 45°F [7°C]) all of the time for the AHU cooling coils.

Air-Handling Units
The AHUs are provided with demand control ventilation. CO_2 sensors in the space monitor CO_2 levels and modulate the outside air supply rate by either direct activation of the variable frequency drive (VFD) speed or actuation of the economizer dampers, depending on the configuration of the AHU.

When the space load can be met by direct outside air the coils are bypassed by the actuation of the bypass damper.

Lessons Learned from HE.5
HE.5 systems design focused primarily on reducing overall annual energy use (kilowatt hours) and successfully reduced the number of compressor hours through the combination of evaporative cooling and use of the active/passive radiant floor.

The integration of radiant floor with minimum fresh air ventilation had two primary benefits:

- Reduced fan energy
- Reduced air-side load due to limited convective heat transfer from the external envelope.

SUMMARY OF PROGRESS TOWARD A SUSTAINABLE BUSINESS MODEL

Walmart's drive to reach a sustainable business model continues. The approach they have taken is transferrable to most large organizations and can be summarized as follows:

- Clear direction from the boardroom
- Task-specific groups set up with the authority to make changes
- Independent measurement of results.

It has been far from the straight-line improvement envisioned in Figure 14.1. Probably one of the key lessons is that to change the direction and performance of such a large organization takes stamina. The organization needs to recognize that it is in this for the long haul and there will be measures that fail, but if it learns from the failures as well as the successes progress will be made. Arup has been working with Walmart for nine years now and we continue to investigate improvements, such as the benefits of green roofs over white roofs, modular PV shade systems, and improved vestibule design.

When we focus on the energy performance of the stores, there are several lessons:

- Some of the systems developed to reduce energy are far more sophisticated than is typical for the big-box construction sector. A lot of effort is needed in commissioning and checking the controls to make sure the predicted performance is achieved in practice.
- Keep looking for ways to simplify the systems. This will reduce first costs and help to avoid the issues mentioned in the previous point.
- Walmart have used their position to drive innovation in energy products. These include simpler radiant pipe systems (see the box "Using Walmart's Purchasing Power to Drive Innovation"), LED produce and parking-lot lighting, and more efficient rooftop package AHUs, among others.
- As a market leader they force other big retailers to follow. Today nearly all the major big-box retailers have sustainability and energy reduction programs underway.

As designers, when we have been faced with owners who say they can't afford the sustainable option as a business, the Walmart story is a strong counterargument. There are few who can argue that Walmart is not financially robust and bottom-line focused, but it is fully invested in making its operations sustainable.

NOTES

1. See https://www.cdproject.net/.
2. October 24, 2005; quoted in Walmart presentation.
3. Walmart, "2011 Global Responsibility Report," Bentonville, AR: Walmart, 2011; see http://walmartstores.com/sites/ResponsibilityReport/2011/sustainable_overview.aspx.

PART 3

Adaptation Strategies

San Francisco's Embarcadero during a 2011 "King Tide" event—the highest tide of the year
Image: Sergio Ruiz

15

Introduction to Adaptation and Resilience

Cole Roberts

> Water, water, every where,
> Nor any drop to drink.
> Samuel Taylor Coleridge

These oft-quoted lines from *The Rime of the Ancient Mariner* have a ring of truth when considering some of the climate changes predicted for this century. Rising sea levels will create problems for low-lying coastal communities but do nothing for areas that could see extended drought. Although precipitation will increase in some areas, it is likely to come in more intense bursts with consequent flooding.

THE CHALLENGE

Imagine a revolutionary technology for energy generation that emits no harmful pollutants, greenhouse-related or otherwise, and poses no safety or political risks. Imagine that we could build this revolutionary technology into our communities and transportation systems—placing it in operation in a matter of a few days anywhere in the world, providing clean, equitable, and abundant power the world over. What then?

We would still be faced with a changing climate and the need to adapt to it.

Scientists from the U.S. National Oceanic and Atmospheric Administration have reported that changes to global climate due to current greenhouse gas (GHG) emissions will be largely irreversible for 1,000 years after emissions stop.[1] Called *overshoot*, the analogy is that of a projectile fired through the air. Even after the propulsion source that launched it is snuffed out, the projectile continues on its course until it finds a resting place some distance farther on. Stopping the overshoot of the global warming projectile is considered to be an unreasonable expectation, even with transformative technological advances and aggressive investment in carbon sequestration.

Overshoot will occur even if we had a revolutionary technology that we could deploy tomorrow. We do not. The economic challenge of redirecting capital to tackle the problem, the political challenge of bringing diverse stakeholders together, and the technological reality that we have no easy fix means that reducing our overshoot will be difficult even under our current level of emissions. And "business-as-usual" forecasts show an *increase* in emissions, not a decrease.

Even where there is a recognition of climate change risk and a declared intention to act, the results often fall short of the aspirations. The landmark United Nations conference on climate change held in Copenhagen in 2009 set an objective of limiting global temperature rise to 2°C, in keeping with an atmospheric carbon dioxide equivalent (CO_2e) level of 450 parts per million (ppm). But the agreement is nonbinding, like a contract that no one will sign. Even if new policies *are* enacted as a result of the accord, they are likely to be cautious rather than aggressive. Such guarded policy implementation is projected to stabilize emissions not at 450 ppm but at 650 ppm, resulting in a near-doubling of global warming temperatures and compounding the impacts of climate change.

Given that the elderly and impoverished are most vulnerable to the impacts of climate change (due to their reduced capacity to

> **The cost of getting on track to meet the climate goal for 2030 has risen by about $1 trillion compared with the estimated cost in last year's World Energy Outlook.** This is because much stronger efforts, costing considerably more, will be needed after 2020. In the 450 Scenario in this year's Outlook, the additional spending on low-carbon energy technologies (business investment and consumer spending) amounts to nearly $18 trillion (in year-2009 dollars) more than in the Current Policies Scenario, in which no new policies are assumed, in the period 2010–2035. It is around $13.5 trillion more than in the New Policies Scenario.
>
> **The timidity of current commitments has undoubtedly made it less likely that the 2°C goal will be achieved.** Reaching that goal would require a phenomenal policy push by governments worldwide: carbon intensity—the amount of CO_2 emitted per dollar of GDP—would have to fall at twice the rate of 1990–2008 in 2008–2020 and four times faster in 2020–2035. The technology exists today to enable such a change, but such a rate of technological transformation would be unprecedented. These commitments must be interpreted in the strongest way possible with much stronger commitments adopted and acted upon after 2020, if not before.
>
> *World Energy Outlook 2010*[2]

adapt), the lack of progress on the Millennium Development Goals is a further setback.[3] At the beginning of the twenty-first century, the number of people living in absolute poverty is estimated to be approximately 1.7 billion—greater than the number of people living at the beginning of the twentieth century.[4]

As meaningful policy languishes, the trends of population increase, urbanization, and wealth creation in developing countries will drive an increase in global GHG emissions even as efforts to reduce emissions build momentum. Consider that every two months an urban population equivalent in size to the city of Paris is added to the globe.[5]

At the same time, developed countries, which have historically been responsible for most of the world's GHG emissions, have an existing built environment and culture that are not easy to adapt. Even if the behavior of the people could be easily swayed, the shape and density of buildings, the orientation and grid of city blocks, and the dominant modes of transit are already constructed. The resulting physical inertia poses a significant hurdle as civilization strives to reset the status quo, or default condition (discussed in Chapter 11), and to find ways to cost-effectively redevelop existing buildings and communities (discussed in Chapters 7 and 8).

And finally, consider the year 2101. The vast majority of climate change research has focused on an arbitrary but generally agreed upon milestone year of 2100. Of course, we all hope that civilization will continue into the twenty-second century, and with it much of our built infrastructure. Shouldn't we also be preparing for 2101 and beyond?

A CASE FOR OPTIMISM

All of this sounds remarkably concerning, and it should. The cost of adaptation in developing countries alone is projected to approach $100 billion per year.[6] Fear, concern, and guilt have a tendency to spur action. But so does optimism. Throughout history, humanity has been remarkably adaptable. In our efforts to eradicate polio, we saved millions of lives. In competitive spirit, people have landed 1960s-era technology on the moon and have sent spacecraft beyond our solar system to explore the depths of space. Each of our cities and towns reflects a deep and embedded capability for working together to

address complex and interdependent problems. Clearly, if we are motivated to act and effective in our actions, we can adapt.

A successfully adapted built environment will ensure that money is invested proactively rather than spent reactively. It will ensure our communities are resilient to disasters and not hobbled by them. Businesses will be quick to reopen. Loss of life will be minimized. And government will be able to maintain services and security.

The benefits of successful adaptation will go beyond addressing climate change risks. An adapted built environment will grow our global and national economies,[7] allow for more effective post-disaster response to threats such as earthquakes and terrorism, and improve the quality of life for the older members of our communities, women,[8] children, and the economically impoverished.[9]

Every day, communities invest time, money, and heart to improve their built environment. Ensuring that these investments are smart is our greatest opportunity for adapting successfully.

UNDERSTANDING ADAPTATION

History teaches us valuable lessons in successful adaptation—and conversely in how failure can sneak up on us. The noted scientist and author, Jared Diamond, illustrates a three-stage process that occurs in any society facing threat.[10] Failure at any stage has a historic precedent of collapse and dissolution of otherwise successful civilizations.

1. *Recognize* A society must perceive a threat.
2. *Choose to act* A society that perceives a threat must then choose to avoid or adapt to it.
3. *Act successfully* A society that chooses to act is not enough. The action must be effective and timely.

The framework is as powerful and robust as it is simple. It is drawn from examples of civilizations as varied and successful as the Roman Empire, Easter Island/Polynesia, Montana's logging communities, and modern-day China. Collectively, our global community is in all three stages at once. Some of us are failing to recognize the implications of climate change. Others recognize it, but are choosing not to act. Others are acting, but not effectively. And perhaps only a few are acting successfully—and their successful actions are not easy to discern.

Uncertainty and Risk

The climate is remarkably complex, and only in the past fifteen years has the computational capacity of supercomputing reliably demonstrated an ability to model its multitude of interdependent systems. Adding economic and social complexity to climate modeling results in significant additional uncertainties. Uncertainty is no justification for inaction. An effective adaptation response recognizes the uncertainty in scenario outcomes and takes a similar approach to financial investing—diversification, scenario planning, and risk/reward decision making. Denying the necessity to act is analogous to putting savings under the mattress; it only feels smart in the short term. Instead, thoughtful investment backed up by available data is the responsible course for risk-management planning. On balance, those who invest gain from the investment.

In addition to basic understanding of risks, a key to understanding climate change risk is the accumulation of risk created by compounding events linked to interconnected systems—termed *concatenated hazards* by Allan Lavell.[11] Primary hazards lead to secondary hazards, which lead to additional hazards in a domino effect. Urban development, especially insufficient or inadequate urban development, tends to be especially at risk of such events given the population density and inherent interdependencies of a city or region. However, the increased risk is often balanced by the greater ability and lower cost per capita of adapting to such threats, as a result of greater organizational efficiency, emergency preparedness, and economic strength.

TABLE 15.1
Understanding disasters and their relationship to risks
Risk continuum adapted from work by Bull-Kamanga, et al., 2003[12]

Event	Disasters	Small disasters	Everyday risks
Occurrence	Infrequent	Frequent (seasonal)	Daily
Severity	10+ killed 100+ seriously injured	3–9 killed 10+ injured	1–2 killed 1–9 injured
Impact (potential)	Can be catastrophic but low impact overall	Significant and underestimated	Main cause of death and injury
Risk continuum	High severity Low frequency ◄——————————————————►		Low severity High frequency

Because the framework of societal collapse is the same as for societal success, it can be used to outline a framework for understanding adaptation.

Recognize: Top-Down and Scenario-Driven

The development of macro-scale scenarios often becomes a catalyst for increasing consciousness of climate change. These efforts are typically in the form of national government policy and global climate modeling, where results are subsequently downscaled to regional levels and communicated in the form of potential scenarios. Such top-down, scenario-driven approaches have been shown effective in answering questions such as "What are the key long-term impacts of climate change?" and "To what extent can the harmful effects of climate change be reduced through adaptation?"[13] The scenarios outline

impacts in a range of areas, including the built environment, where direct and indirect scenario-based impacts include the following:

- *Change in extremes (direct)*[14] Many of the places that we choose to build or have chosen to build in the past face new hazards. Among the effects of global warming is an increase in extreme weather events, including:
  - Hurricanes and associated storm surges
  - Floods (from more intense rain events or accelerated thaw of snowpack)
  - Tornados.
- *Change in averages (direct)* There will be changes in normal weather patterns and climate conditions that affect the day-to-day performance of our buildings and infrastructure, including:

- Average and design temperatures (heat, cold, and humidity)
- Design wind speeds
- Average precipitation
- Average sea levels.
- *Indirect change* Some of the most severe impacts to human quality of life are not directly a result of extreme or average changes in climate, but rather occur as an indirect outcome. Examples include:
  - Global and national economic stress from disaster response spending
  - Economic stress due to failed/declining industries (e.g., fisheries, tourism, agriculture)
  - Global and national political destabilization due to increasing stress on impoverished populous communities (e.g., increased piracy, border conflict, and terrorism)
  - Ecological services collapse or decline (e.g., termite damage, bark beetle damage, wheat rust, wetland ecosystem failure, barrier reef failure).

The earthquake that struck Japan in the spring of 2011 underscores the direct and indirect multidimensionality of the threat of hazards. The greatest impacts come not from a single event, but from the confluence of multiple stressors in sequence or simultaneously, and the resulting coping capacity of those affected. Understanding such risk accumulation is critical to *full* recognition of climate change threats and the incentive to act on them.

Choose to Act: Bottom-Up and Vulnerability-Driven

A top-down, scenario-driven approach from government or institutions can provide much-needed context for and understanding of the potential impacts of climate change in the built environment; however, this perspective has proved generally inadequate in committing people and organizations to local action.

In order to cement such a commitment to action, an understanding of localized vulnerability and personal relevancy is needed within communities and industries.[15] In the built environment, this means a clear assessment of the direct impacts on built infrastructure (e.g., flooding, fire danger) and the indirect impacts on the community (e.g., loss of local jobs and tourism dollars). Such an assessment must

also highlight the local capacity to adapt. Since vulnerability is a product of (a) risk and (b) ability to cope (see the box "Understanding Vulnerability"), this capacity is often markedly different between groups in the community. Large financial and emotional vulnerabilities become evident to those who own property or are culturally vested in the area impacted, regardless of their income. Significant health and livelihood vulnerabilities are often present in impoverished and elderly communities. The approach to adaptation will vary based on each group's ability to organize and affect the priorities of the greater whole.

Although some stakeholders often have little power individually, there is ample historical proof that when the powerful fail to share their *stability* with the powerless, the powerless share their *instability* with the powerful.

Act Successfully: A Comprehensive and Time-Based Approach to Adaptation

Creating a built environment that has the ability to cope with climate change will take decades and perhaps centuries of sustained effort. Although the decisions made today will shape the services and cultural richness of our children's future, such a long time horizon makes it very difficult to act successfully. The value of future benefits is simply devalued (i.e., *discounted*) the further into the future they are promised.

Successful action therefore needs to exist within a framework that is (a) comprehensive (i.e., inclusive of all adaptation actions), (b) prioritized, and (c) time-based, recognizing the iterative life cycle of planning, building, and infrastructure. In the guideline chapters that follow, the authors introduce an approach that achieves these requirements and has been used for effective action.

ADAPTATION ROLES

As in any period of change, there are opportunities. The opportunities to participate *and benefit* from climate change adaptation are open to many elements of society, according to their skills and authority.

Government is positioned as a local, national, and global authority able to invest, legislate, and enforce. Political advantage will be gained in the trenches of inspiration, fear, and financial contribution.

Understanding Vulnerability

There are a variety of definitions and approaches to understanding vulnerability.[16] The authors recommend the Vulnerability of Places approach, which links the distribution of exposure to natural hazards and the distribution of a population's coping ability across a geographic area.[17] The strength of this approach is that it is multilayered. It recognizes the variety of hazards that may present risks. It also recognizes that vulnerability will differ based on the makeup of the community and its exposure to the hazards. Within a community, coping ability will vary by income, wealth, gender, age, health, family structure, and house ownership. An analogous distribution may occur within organizations, where coping ability will vary internally by staffing levels, structural hierarchy, language, culture, financial resources, etc.

　　This approach is defined for practical application as
V = S × O × C, relative to a geographic area.

Vulnerability = Severity of a hazard × Occurrence likelihood × Coping *in*-ability

given that:

Vulnerability = Risk × Coping *in*-ability

(the higher the risk, the higher the vulnerability; the poorer the coping ability, the higher the vulnerability)

Risk = Severity of a hazard × Occurrence likelihood

Coping *in*-ability = inverse of coping ability

Coping ability (aka adaptive capacity) = Resilience (the ability to *recover* from the damage caused by a hazard) + Resistance (the ability to *withstand* the damaging effects of a hazard)

　　Effectively, vulnerability is the inverse of resilience when all else is held equal. This idea is common in many discussions that emphasize improving resilience in order to lessen vulnerability.

15.1 Elements of vulnerability
Graphic: Arup/Cole Roberts

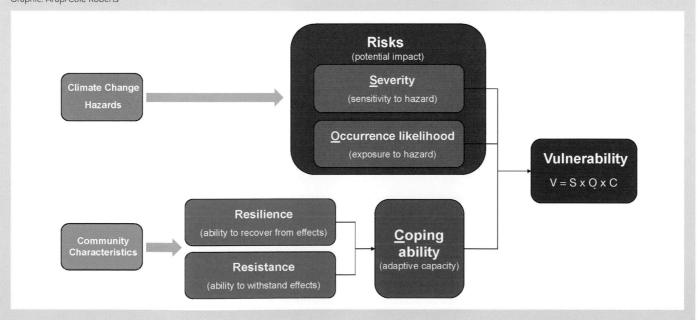

Increased federal emergency management funding will become increasingly important in disaster response. Large public works projects that protect diverse community interests will be needed. Legislative requirements for resilient infrastructure and planning will change from standards based on historical monitoring to standards based on predictive futures modeling. Martial authority will be used to support mandatory evacuation orders. Martial authority will also be used to manage rioting as services are increasingly compromised and quality-of-life changes stir discontent.

Industry is especially well positioned to create profit opportunities out of the risk, and capture market share through effective management. Adaptation will require significant professional service, manufacturing, and construction industry participation. As our new built environment takes shape and our existing one is rebuilt, jobs will grow. Changes in land value, insurability, and infrastructure investment will create winners and losers.

Academia will serve in partnership with national laboratories to inform national policy and industry direction. It will also prompt conversation to go beyond recognition and toward a debate over "What is the right action?"

Mission-driven organizations and individuals within communities of purpose will be able to leverage climate change adaptation by linking it to their core mission. Examples of organizations that have the potential to be more effective and in greater demand include social justice, general education, family, and pro- and anti-immigration rights groups. The reasons are as diverse as the organizations. Social justice organizations will be in greater demand as vulnerable populations suffer increasing disadvantage when access to resources decreases and threats increase. Governments under economic pressure to reduce educational services will increasingly rely on and be spurred into action by general and specialist education organizations. Immigration will intensify as direct and indirect impacts of climate change create dislocations of people. As the ability to cope is strained, family services organizations will increasingly be called on for support.

The greatest opportunities will fall to individuals who catalyze action in their respective organizations, government, and industry. Additionally, the collective social network is likely to create opportunities through populist movements where few otherwise exist. Over the years ahead, each person in our society will be faced with challenges of recognition, choice, and right action that will shape culture and behavior.

Those governments, industries, organizations, and individuals with significant vulnerability and limited capacity to respond *will* suffer from climate change. They will include our respected institutions, our aging parents, our economically distressed youth, our impoverished, *and* our powerful. How much we suffer or succeed will depend on our actions. And our choice of actions is our own.

ADAPTATION IN SUMMARY

Changes to global climate due to current GHG emissions are largely irreversible for 1,000 years after emissions stop. And as meaningful policy languishes, many trends will drive an increase in global GHG emissions, not a decrease. At the same time, we have an existing built environment and culture that are not easy to adapt.

A successfully adapted built environment will ensure that money is invested proactively rather than spent reactively. It will ensure our communities are resilient to disasters and not hobbled by them. In the adaptation of the built environment, direct and indirect changes will occur, with indirect impacts most likely to affect daily quality of life. Although there is uncertainty in the economic and social impacts, that uncertainty must be managed.

One of the keys to understanding climate change risk is the risk accumulation created by compounding events. Primary hazards lead to secondary hazards, which lead to additional hazards, in a domino effect. Urban development tends to be especially at risk of such events. However, increased risk in these areas is often balanced by greater ability and lower cost per capita to adapt to such threats.

The best inspiration comes from historical lessons in successful adaptation—and conversely in how failure can sneak up on even the strongest societies. Societies must recognize a threat when it is present. They must then choose to avoid or adapt to it. And finally, their actions must be effective.

In order to recognize the potential impacts of climate change, top-down, scenario-driven approaches are most effective. In order to catalyze the choice to act, bottom-up, vulnerability-driven approaches are needed for personal relevancy. And finally, successful action needs to exist within a framework that is (a) comprehensive, (b) prioritized, and (c) time-based, recognizing the life cycle of planning, building, and infrastructure.

As in any period of change, there are opportunities. Governments, industries, organizations, academia, and individuals will stand to benefit from successful adaptation. Those with significant vulnerability and limited coping capacity *will* suffer. They will include our respected institutions, our aging parents, our economically distressed youth, our impoverished, *and* our powerful. How much we succeed will depend on the choices we make.

NOTES

1 Susan Solomon, Gian-Kasper Plattner, Reto Knutti, and Pierre Friedlingstein, "Irreversible Climate Change Due to Carbon Dioxide Emissions," *Proceedings of the National Academy of Sciences of the United States of America* 104(6), 2009: 1704–9.

2 International Energy Agency, 2010, http://www.iea.org/weo/2010.asp (accessed June 23, 2012).

3 Millennium Development Goal reports can be accessed at http://www.un.org/millenniumgoals/reports.shtml.

4 Graziella Caselli, Gillaume Wunsch, and Jacques Vallin, *Demography: Analysis and Synthesis: A Treatise in Population*, Burlington, MA: Academic Press, 2005.

5 Approximately 77 million persons are being added annually (Population Division, UN Department of Economic and Social Affairs, *World Population Prospects: The 2010 Revision: Highlights and Advanced Tables*, New York: United Nations, 2011). The Paris metropolitan area population is approximately 12 million.

6 Economics of Adaptation to Climate Change Study team, *The Cost to Developing Countries of Adapting to Climate Change*, Washington, DC: World Bank, 2010.

7 Stern Review: The Economics of Climate Change, 2006, http://webarchive.nationalarchives.gov.uk/+/http:/www.hm-treasury.gov.uk/independent_reviews/stern_review_economics_climate_change/stern_review_report.cfm (accessed June 5, 2012).

8 Women are included as they may have needs that are not well served in communities under stress. They may be disadvantaged as a result of role (primary/sole caregiver), health status (pregnancy), income (lower pay or poorer job access), mobility (right to drive), authority (choice to relocate), cultural/religious expectations (clothing restriction), etc.

9 David Satterthwaite, Saleemul Huq, Mark Pelling, Hannah Reid, and Patricia Romero Lankao, *Adapting to Climate Change in Urban Areas: The Possibilities and Constraints in Low- and Middle-Income Nations*, London: International Institute for Environment and Development, 2007.

10 Jared Diamond, *Collapse: How Societies Choose to Fail or Succeed*, New York: Viking, 2004. Diamond actually identified four categories of response to societal threat: (1) Failure to anticipate, (2) Failure to perceive, (3) Failure to act, (4) Failure of solution. We have combined the first two categories.

11 "Natural and Technological Disasters: Capacity Building and Human Resource Development for Disaster Management: Concept Paper" 1999.

12 Liseli Bull-Kamanga, et al., "Urban Development and the Accumulation of Disaster Risk and Other Life-threatening Risks in Africa", *Environment and Urbanization*, 15(1), 2003: 193–204.

13 United Nations, "Application of Methods and Tools for Assessing Impacts and Vulnerability, and Developing Adaptation Responses," background paper for the United Nations Framework Convention on Climate Change, Buenos Aires, Argentina, FCCC/SBSTA/2004/INF.13, November 10, 2004.

14 Earthquakes, though not climate-related, are a functional analogy for response to extreme hazards, since significant knowledge of community resilience, disaster response, and insurability has been developed in response to earthquake risk.

15 See note 10 above.

16 Three documented approaches are (1) hazard focus: vulnerability due to distribution of hazardous conditions and how it affects people and structures (i.e., the likelihood of events occurring and the estimated damage that will result), (2) coping distribution: vulnerability due to distribution of resistance and resilience between different elements of a populace, assuming a given hazard is constant, and (3) place-based: vulnerability due to distribution of hazardous conditions *and* coping ability within a geographic area.

17 S. L. Cutter, "Vulnerability to Environmental Hazards," *Progress in Human Geography* 20(4), 1996: 529–39; S. L. Cutter, J. T. Mitchell, and M. S. Scott, "Revealing the Vulnerability of People and Places: A Case Study of Georgetown County, South Carolina," *Annals of the Association of American Geographers* 90, 2000: 713–37; Shuang-Ye Wu, Brent Yarnal, and Ann Fisher, "Vulnerability of Coastal Communities to Sea-Level Rise: A Case Study of Cape May County, New Jersey, USA," *Climate Research* 22, 2002: 255–70.

The Thames Barrier is the world's second largest movable flood barrier (after the Oosterscheldekering in the Netherlands) and is located downstream of Central London
Image: Iain McGillivray/Shutterstock.com

16

Planning for Adaptation and Resilience

Cole Roberts

The reasonable man adapts himself to the world; the unreasonable one persists in trying to adapt the world to himself. Therefore all progress depends on the unreasonable man.[1]

> George Bernard Shaw, *Man and Superman*, 1903

This chapter explores the climate change predictions, risks, and guidelines for successful adaptation. It builds on Chapter 15's introduction to adaptation by providing guidance on how to effectively plan, design, and make policy. It also serves as a generalized introduction to the regional guideline chapters that follow, which explore adaptation in sequentially greater detail and design relevance for select regions.

Although this chapter and its paired regional chapters capture the necessary actions for successful adaptation, they are limited in their ability to generalize *vulnerability*, the combination of risk and adaptive capacity. Although the risks are reasonably consistent between similar climatic regions of the world, the adaptive capacity is not. Even where climates are similar, differences in culture, wealth, government, and social services mean that the gulf between the least and most vulnerable regions is wide. Although not all of the regional chapters will be relevant to all regions, the content serves as a starting point for understanding the adaptation response. As you read, consider how your region compares, imagine the people you know choosing (or not choosing) to act, and imagine the plausible futures that are even now being shaped.

RISK ACCUMULATION

All regions exhibit a relatively unique potential for risk accumulation due to their unique characteristics (infrastructure, social organization, exposure, etc.). In any adaptation planning effort, it is this potential that should be understood in order to gauge progress. And it is often this accumulated risk that provides significant opportunity for an integrated adaptation response. Each of the chapters following this highlights examples of such risk accumulation issues.

Integrated Adaptation

The business-as-usual "predict and prevent" (P&P) response to vulnerability (e.g., increased flooding predictions demand higher roads) is no longer viable when addressing significant uncertainty and the risk accumulation of urban systems. The reasons are twofold. First, such ways of thinking lead to false security in "fail-safe" systems, such as the levees that "protected" New Orleans, instead of "safe failure" strategies consistent with healthy wetland systems and emergency response investment. Second, P&P has the potential to lead to maladaptation: the increasing of vulnerability elsewhere in the city or community through a narrow focus on individual "climate-proofing" projects. A key insight of work undertaken in ten cities in Asia (the Asian Cities Climate Change Resilience Network) has been the emergence of an integrated adaptation approach that includes both "soft" and "hard" interventions as the only path that avoids maladaptation and potential catastrophic failures stemming from the unpredictability of climate change risks and their interactions. In addition, early engagement with "softer" measures like integrated planning and policy was identified as critical to avoiding a trajectory of costly, maladaptive urban development. The process encourages flexible and integrated infrastructure systems that advance more than one aim. To capture the integration opportunities, holistic design and planning occurs in the earliest phases of development: "Once conceptual designs begin, the prospects for incorporating effective adaptation and resilience building measures dips dramatically."[2] (See also Chapter 10.)

RECOGNIZE: ANTICIPATED CHANGES

Scenario-based modeling of the changes likely in a region is summarized in the chapters that follow. The figures are drawn from data sourced from the World Bank Climate Change Knowledge Portal, an online data portal that includes information from nearly

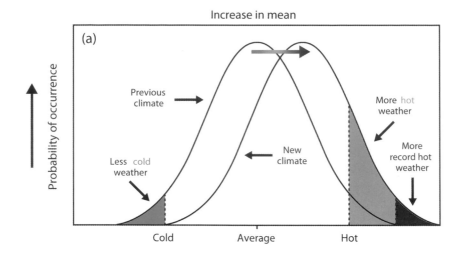

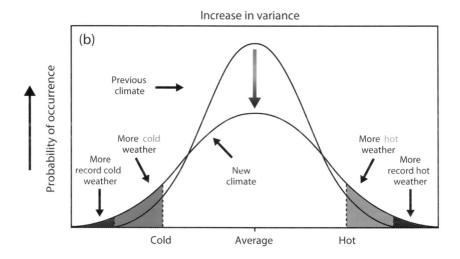

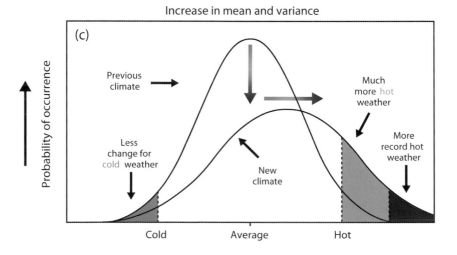

16.1 Direct changes in weather occur both in the extreme (variance) and in the average (mean)

Graphic: IPCC, Solomon et al., 2007 [i]

two dozen climate change models prepared by researchers around the world.[3] These primarily "top-down" scenario-based models can prompt recognition of climate change but are not typically adequate to spur local action because they don't fully account for localized vulnerability.[4]

CHOOSE TO ACT:
VULNERABILITY IN THE BUILT ENVIRONMENT

Bottom-up, vulnerability-derived action is often limited by insufficient local data, organizational structure, and social momentum. There is an understanding that something bad is coming, but the details are murky. As a result, when a community of leaders first chooses to act to improve the resilience of its built environment, successful action is simply setting a framework for near- and long-term action and establishing a process that will change over the ensuing years and decades.

An outline for an adaptation plan may include the following:

1. *Climate change overview (top-down scenarios)* What's happening globally and nationally?
2. *Local concerns (bottom-up vulnerabilities)* What does this mean for us?
3. *Convening of stakeholders* Who should be participating?
4. *Establishment of focus* Goal setting, strategy, and information needed.
5. *Integrated adaptation* Resistance measures (reflecting the ability to withstand damaging effects) and resilience measures (reflecting the ability to recover from damage caused).
6. *Timeline and process* Near-term and long-term activity.

Moving from a commitment to action to ongoing action requires a sustained and reasonably comprehensive effort based on the principles of successful decision making over years and generations. Such principles include many of those discussed in earlier chapters:

- *Behavior change must be supported to be sustained (Chapter 11)* Successful choice is a product of (a) motivation, (b) ability, and (c) triggers to act (C = MAT).

- *Natural disaster is a failure in artificial development* That is, a lack of adequate ability to cope with a natural occurrence.
- *Don't aim for zero impact* Adaptation is about reducing the impact and improving the quality of response.
- *Rationalization is not adequate* Often the irrational argument based in our emotional intelligence is stronger than the rational argument.
- *Find comfort in unpredictability* Uncertainty is not a justification for inaction.
- *Seek "safe failure" outcomes that avoid catastrophic loss*
- *Integrate adaptation* Solve multiple challenges with single actions.
- *Learn* Mistakes will be made, but they are needed to progress.
- *Seek leverage* Multiply the value of investment by capturing cobenefits of development.

One of the most challenging aspects facing governments and developers is the uncertainty of knowing what local impacts can be expected, when impacts are likely to be felt, and how severe these will be. To make matters more complicated, cities and developers have been presented with a considerable obstacle: how to begin to address planning and investment for climate change resilience in difficult economic times and consequent fiscal distress.

This predicament opens up opportunity for innovative financing and public–private partnerships. The Global Adaptation Institute is one organization working to bring the public sector and private investors together and accelerate investment in climate change adaptation. Founded by Dr. Juan José Daboub, former managing director of the World Bank (2006–10), the institute is working to help open a market in investment in climate change adaptation to create opportunities for engaging the business community.

Vulnerability Mapping and Assessment

In order to better understand a community's vulnerability, vulnerability mapping is needed as an element of any adaptation plan. It should accompany a summary of the risks and coping ability as a complete assessment. In each of the chapters that follow a Vulnerability Map is included along with a brief narrative summary of a variety of topics that

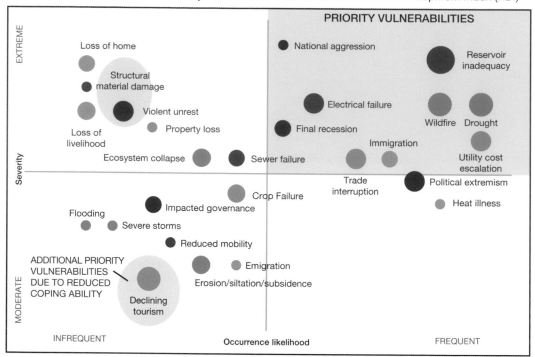

Example HOT AND DRY Community with HIGH to VERY HIGH UN Human Development Index (HDI)

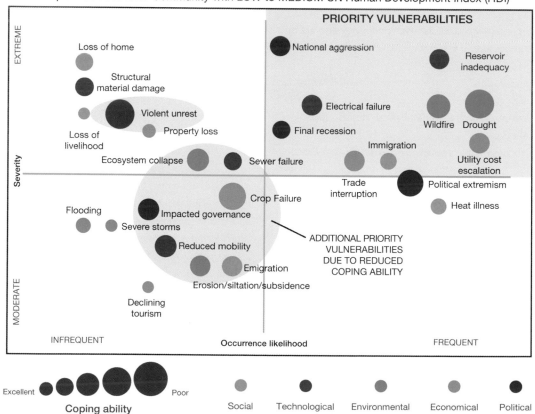

Example HOT AND DRY Community with LOW to MEDIUM UN Human Development Index (HDI)

16.2 An example Vulnerability Map with highlighted priority areas

Graphic: Arup

have potential to impact the built environment in the particular region. We include an example here (Figure 16.2) to introduce the concept. It is a criteria-based map (as discussed in Chapter 11) that scores risks on a two-dimensional axis. Additional criteria—coping ability, and categorization of risks as social, technological, environmental, economic, and political—are used to size and color the plotted points to convey vulnerability and interrelationship.

The example map uses the United Nations Human Development Index (HDI) as an arbitrary index to suggest coping ability. Although urban and rural areas in high-HDI countries may have advantages due to income or governance structure, the use of HDI or any generalized index for coping ability is not recommended in practice. Actual vulnerability will vary significantly within a community, and it is critical that assessments are of adequate rigor to account for such differences. Consider, for example, the significant vulnerability of individual neighborhoods and populations within New Orleans, Louisiana, during Hurricane Katrina in 2005.

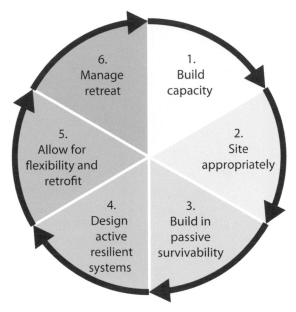

16.3 A six-step adaptation approach that is comprehensive, prioritized, and time-based
Graphic: Arup

ACT SUCCESSFULLY: THE DESIGN, PLANNING, AND POLICY RESPONSE

Although the climate change risks may be severe for a region, there are significant opportunities for implementing responses to the risks that lessen vulnerability. Further, there are multiple cobenefits of these responses. Among them is the potential for increased job creation, economic resilience, and sustained cultural heritage. In each of the regional adaptation chapters that follow, we have organized the design response guidelines into a six-step adaptation approach (Figure 16.3). The intent is to achieve comprehensiveness, prioritized cost effectiveness (from most to least cost effective), and timeliness of action.

Since many of the recommendations stay with buildings and urban systems for their lifetime (e.g., building massing and street form), they are recommended for immediate action. Indeed, many of these solutions can be seen in current "green" building and sustainable planning efforts throughout the world. Other systems have shorter lifetimes and are therefore noted as changes that can be deferred (e.g., building air-conditioning equipment). Other recommendations will take years to enact and require a consistent effort that gradually

increases in intensity over time (e.g., managed retreat or urban building retrofit).

The recommended responses are, in keeping with this book's subject, *relevant only to the built environment*, and do not include adaptation responses that are unrelated to the built environment but are no less important (e.g., under "Build Capacity," increasing education and social services is not discussed).

1. Build Capacity—Institutional and Technical (Immediate Action)

First steps are the most valuable. Start to build capacity in the local community, institutions, and technical practice. The emphasis should be innovation, value-based planning, and the establishment of a continually improving process that challenges the default condition. For example:

An Urban Resilience Framework

An Urban Resilience Framework (URF) has been developed by Arup and the Institute for Social and Environmental Transition based on research, academic work, and experiences as practitioners, including collaboration on the Rockefeller Foundation's Asian Cities Climate Change Resilience Network. Its intent is to provide an umbrella framework for improving urban resilience. Unlike many approaches, the URF is grounded in the interdependencies of urban systems and a process of evolutionary learning over time. A simplified version of the URF has been integrated into this chapter. Readers are encouraged to access the original and complete URF, which includes additional detail, notably on the roles and relationships between internal and external actors, urban system relationships, and a formalized relationship with nonprofit and government funders.[5]

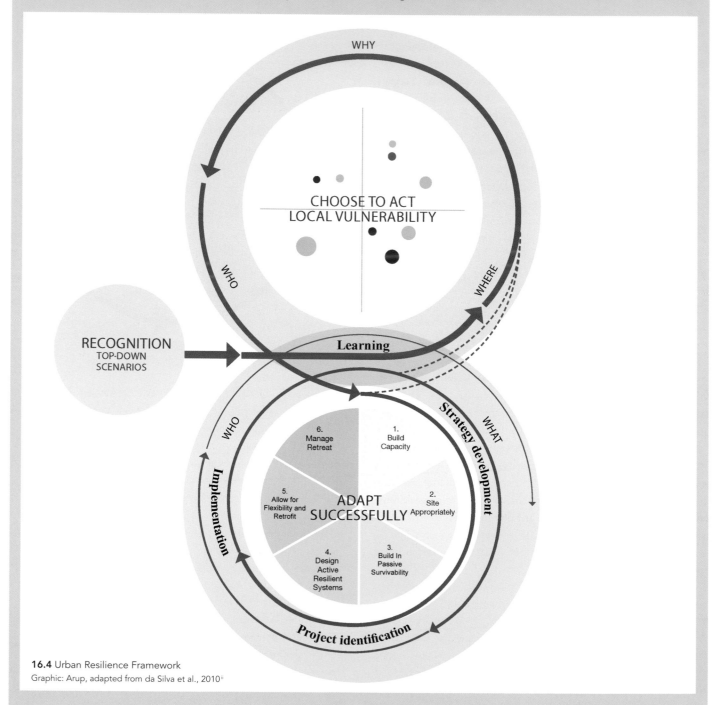

16.4 Urban Resilience Framework
Graphic: Arup, adapted from da Silva et al., 2010[ii]

- *Social* Establish convening organizations for improved integrated decision making, planning, and design.
- *Technological* Support research that creates needed knowledge for improved professional practice.
- *Economic* Create innovative financing mechanisms that lessen first cost hurdles for owners and amortize adaptation investments over time.
- *Environmental* Build capacity in the insurance industry to value climate change risk and ecosystem service protection.
- *Political* Promote a reset of the default condition to economic innovation and improved quality of life for voters and their families.

2. Select the Right Site—Reduce Exposure
(Immediate Action)

Construct buildings and infrastructure on sites where their lifespan (or depreciated value) is within the time horizon of the site's viability (e.g., a 100-year structure should not be constructed within an area of 100-year flooding, *adjusted for climate change variation*). Where there is limited supply of unique land quality (e.g., prime farmland), development should be restricted. Siting should support resilient development models, such as infill development, transit-oriented development, and smart growth/smart shrinkage.

According to the saying, "Prevention is better than cure." This is true with city planning—city agencies have a suite of policy and planning tools available to them to reduce the exposure of buildings, infrastructure, and people to climate risk before an intervention is needed. Through efforts to encourage the appropriate use of land and promote construction of built projects on the right site—or conversely, selecting the right development for a site—city governments can make significant steps toward reducing future risk.

In implementation, however, promoting the "right" development for a site can be tricky for policy makers to act on, as changes to codes, plans, or zoning designations may affect existing buildings and the constituents of elected officials, and change the tax base in communities in the short run. But the foresight behind matching development to appropriate sites is intended to reduce the risk of property, business, and personal loss from climate change and increase the economic, social, and environmental health of the community.

3. Build In Passive Survivability
(Immediate Action)

Where buildings and infrastructure are constructed, optimize their ability to survive hazards with reduced or no support (e.g., energy, water, emergency services). This is a sector-based focus (on e.g., transportation, utilities, residential buildings). For example:

- Naturally daylight, ventilate, and temperature-condition buildings through proper orientation, shape, and material choice so they can continue to function without artificial lighting and air-conditioning for much of their occupied time.
- Encourage native and adapted planting and diverse ecological systems for greater resilience to drought and flooding.
- Reduce roof and planting combustibility in fire-prone areas.
- Reduce the use of absorptive materials in flood-prone areas, enabling the built environment to survive occasional partial inundation.
- Use seismic performance objectives as a framework for design survivability criteria following an event—PO 1: immediate occupancy, continued usability; PO 2: life safety, a stable structure but needs finishes and services reworked; PO 3: collapse prevention, requires demolition and reconstruction.[6]

4. Design Active Resilient Systems
(25+ years)

Once passive solutions are expended, construct active systems that are capable of sustaining damage and maintaining partial service. This represents a focus on systemic adaptation—improving networked infrastructure. For example:

- Plan multiple energy and water sources so that a single point of failure in one system does not wholly eliminate the supply.
- Develop community-based infrastructure, such as community-based heating and cooling water systems, where redundant equipment and emergency generation can be more cost-effectively purchased and diversity of loads can reduce overall capital cost.

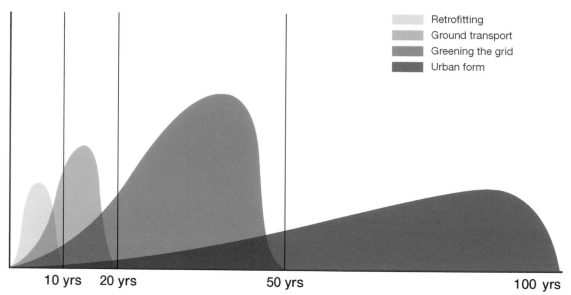

Retrofitting
Ground transport
Greening the grid
Urban form

10 yrs   20 yrs   50 yrs   100 yrs

16.5 Infrastructure and the built environment vary in their ability
to adapt in short amounts of time
Graphic: Arup/John Miles

- Design integrated building and infrastructure systems for future design conditions (e.g., climate change-adjusted design temperature and humidity).[7]
- Increase the temperature of cooling supply air (e.g., to 65°F [18°C] instead of 55°F [13°C]) so that failed chilled-water equipment is less likely to threaten space comfort and induce heat-stress health risks.

5. Encourage Adaptable Buildings and Infrastructure (2+ to 200+ years)

Plan, design, and build for future change by recognizing the adaptation life cycle of built environment "layers." Building fixtures may be changed on a 2–10-year cycle, mechanical and electrical systems on a 10–25-year cycle, envelope on a 20–50-year cycle, structure on a 50–200-year cycle, infrastructure and roadways on a 100–400-year cycle, and cities on a 200–1000-year cycle.[8] Doing so allows for better decision-making rationale, cost-effective retrofit, and life-cycle optimization. For example:

- Make decisions consistent with the time horizon of the layer of built environment:
  - Plan city patterns and boundaries that will make sense to future generations (e.g., don't consume all the local farmland).
  - Build transportation systems that may not be cost-effective for twenty years but will be cost-effective from year 21 to year 100.

  - Bury piping for future use, because it is easy to install today but hard to retrofit later (e.g., dual plumbing, stub cooling connections for future district cooling systems).
  - Design buildings with "good bones"—structures and shapes that can adapt to new skins and new occupants (e.g., parking structures convertible to future retail use, roof forms that allow for future photovoltaic installation).
- Design for systems that can be maintained, disassembled, and replaced at the end of their life cycle.
- Design for effective monitoring and optimization (e.g., smart metering and monitoring systems).

6. Manage Settlement and Retreat (50+ years)

Once all other steps are complete, undertake an assessment of the costs and benefits (both tangible and intangible) of maintaining buildings and infrastructure in geographic areas of our cities and towns that will be severely compromised by the waterscape and landscape changes of the near future. The results of such an assessment may challenge status quo growth assumptions, resulting in a mix of managed settlement and retreat activity:

- *Managed settlement* As the natural and artificial carrying capacity of a region is approached, a variety of strategies may be used to manage growth while maintaining a strong economy. They include urban growth boundaries, development impact fees, and permit restrictions.

- *Passive retreat* As stresses increase, some resettlement will occur through market mechanisms. Commodity costs and rationing will increase and stimulate changes in culture and behavior that will result in individual relocation choices.
- *Active retreat* If it is determined that select areas of the built environment must be sacrificed, the decision will likely require multigenerational notification of those impacted so that people in affected areas are able to gradually divest themselves. The degree of support will vary by locale but may include tax incentives, financial compensation, and partial or full in-kind replacement.
  - Moving infrastructure may include the realignment of roadways, stormwater pump stations, electrical distribution systems, coastal wastewater treatment plants, and coastal nuclear and conventional power stations.
  - Moving buildings may involve temporary protective efforts (e.g., coastal levees) being maintained until the building life has been significantly exhausted, followed by partial or complete demolition and ensuing cessation of the protective efforts (e.g., breaching levees to create new coastlines).

NOTES

1 This quote was chosen because it is provoking—irrational, discriminatory, historical, prescient, and questions the path of progress.

2 Anna Brown and Sam Kernaghan, "Beyond Climate Proofing: Taking an Integrated Approach to Building Climate Resilience in Asian Cities," *UGEC Viewpoints* 6, 2011: 4–7, http://www.rockefellerfoundation.org/uploads/files/cea3570b-01ff-467c-94d8-d80f1fdde9c9-viewpointsvi.pdf (accessed June 11, 2012).

3 http://sdwebx.worldbank.org/climateportal/.

4 The World Bank Climate Change Knowledge Portal is also a good source repository for studies on localized vulnerability, in addition to climatic data.

5 Jo da Silva, Marcus Moench, Sam Kernaghan, Andres Luque, and Stephen Tyler, "The Urban Resilience Framework (URF)," conference paper, Rockefeller Foundation's Urban Climate Change Resilience Seminar, Bellagio, 2010; Marcus Moench and Stephen Tyler, "Systems, Agents, Institutions, and Exposure: A Framework for Urban Climate Resilience Planning" in Marcus Moench, Stephen Tyler, and Jessica Lage, eds., *Catalyzing Urban Climate Resilience: Applying Resilience Concepts to Planning Practice in the ACCCRN Program*, Boulder, CO: Institute for Social and Environmental Transition, 2011, http://www.i-s-e-t.org/images/pdfs/02_ISET_CatalyzingUrbanResilience.pdf (accessed June 26, 2012).

6 *Handbook for the Seismic Evaluation of Buildings: A Prestandard*, FEMA 310, http://www.fema.gov/library/viewRecord.do?id=1534 (accessed June 26, 2012); Andy Thompson, "A Strategic Approach to Loss Prevention Engineering," *Risk Management Magazine*, 2005.

7 Mark F. Jentsch, AbuBakr S. Bahaj, and Patrick A. B. James, "Climate Change Future Proofing of Buildings: Generation and Assessment of Building Simulation Weather Files," *Energy and Buildings*, 40(12), 2008: 2148–68.

8 The list of built environment layers is adapted from Stewart Brand, *How Buildings Learn: What Happens After They're Built*, New York: Penguin Books, 1995. Brand identified six layers: site, structure (30–300 years), skin (20 years), services (7–15 years), space plan (3–30 years), stuff (daily). An adaptive building must allow slippage between each layer.

i S. Solomon, D. Qin, M. Manning, Z. Chen, M. Marquis, K. B. Averyt, M. Tignor, and H. L. Miller (eds.), *Climate Change 2007: The Physical Science Basis: Working Group I Contribution to the Fourth Assessment Report of the Intergovernmental Panel on Climate Change*, New York: Cambridge University Press, 2007.

ii Jo da Silva, Marcus Moench, Sam Kernaghan, Andres Luque, and Stephen Tyler, "The Urban Resilience Framework (URF)," conference paper, Rockefeller Foundation's Urban Climate Change Resilience Seminar, Bellagio, 2010.

Pests and vector-borne diseases are expected to increase as temperatures change, exposing previously untouched communities and structures to damage
Image: mrfiza/Shutterstock.com

17

GUIDELINE CHAPTER

Designing for Warmer and Wetter Climates

Amy Leitch and Cole Roberts

"We are trying to find another place to go, because all the land back home is dissolving," said Mukhles Rahman, who works as a security guard at a garment factory. "But there aren't jobs in other cities or in villages."

Joanna Kakissis, writing in the *New York Times*, describing climate migration in Bangladesh resulting from extreme flooding[1]

The morning is already hot when you leave home, the Sun blazing down from a bright blue sky. Air that smells of asphalt and heat is so thick with humidity that by the afternoon you think you could lift off your feet and swim. Dark clouds gather and build giant storm cells that black out the sky. Birds and insects that earlier were making morning calls fall silent as the pressure builds. Everything is still, waiting. Then, when you can almost imagine the atmosphere pressing into you from above, leaves on the trees begin to shake and curl as the storm approaches. The crack of thunder announces the rainstorm that releases the tension and rain pounds into the ground. Steam rises off the pavement, curling into the air. The stale, heavy air washes away to reveal a fresh new world. With an almost audible sigh of relief, the birds and the insects begin their calls again, greeting the evening. The temperature climbs, and in the morning, the air is hot again.

WATER, WATER EVERYWHERE

Although it receives less attention than its peers, water vapor is the most abundant greenhouse gas (GHG) in the atmosphere.

Generalized, the temperature/water feedback loop describes how, as atmospheric temperature rises, more water vapor is evaporated from "ground storage" (rivers, oceans, reservoirs, soils). Warmer air can hold more water than cooler air, and as a GHG, more water vapor in the air is able to warm the atmosphere, amplifying the cycle.

IT'S GETTING WARMER

In looking at global climate models (Chapter 1), nearly every location is predicted to experience some degree of increased temperatures with climate change, some much more than others (Figure 17.1, overleaf). However, some areas are predicted to experience dramatically reduced precipitation while others are predicted to experience significantly more (Figure 17.2, overleaf).

While local climate change impacts will vary, regions that are likely to experience both warmer and wetter climates include northern latitudes of the United States, Canada and Russia, equatorial regions in Latin America and Asia Pacific, and Antarctica.

About 40 percent of the world's population lives in tropical climates. Developing nations comprise the vast majority of these tropical nations, many of which have exploding populations, rapid urban growth, growing economic inequality, and increasing problems with maintaining environmental quality and planning for and delivering adequate municipal services. Even incremental rises in temperature and precipitation can have disastrous effects.

Expand to mid-latitude and temperate climates, and the map encompasses the vast majority of people on the planet. People tend to prefer living in temperate climates. However, with climate change, current conditions in some parts of the world are set to become warmer and wetter. Relying on a degree of change that is "just right" is shortsighted at best and dangerous at worst.

RISK ACCUMULATION

Regional climates that are shifting to warmer and wetter conditions can lead to risk accumulation in the local built environment.

Passive Design Imbalances

New buildings designed to leverage the current climate for passive conditioning may find future climate change outpaces the capacity of the passive systems to maintain comfort. Parallel increases in temperature and humidity will aggravate the issue.

Heat Stress and Power Failure

Rising temperature and humidity will have a significant impact on communities that are already hot and humid, as upper limits are pushed higher, increasing demand for fuel, energy, and water to maintain health. High heat stress events can be exacerbated by power system failures from extreme events or strained electrical infrastructure (as everyone rushes to plug in new air-conditioning). Accumulated risks are greater in communities that have developed their infrastructure

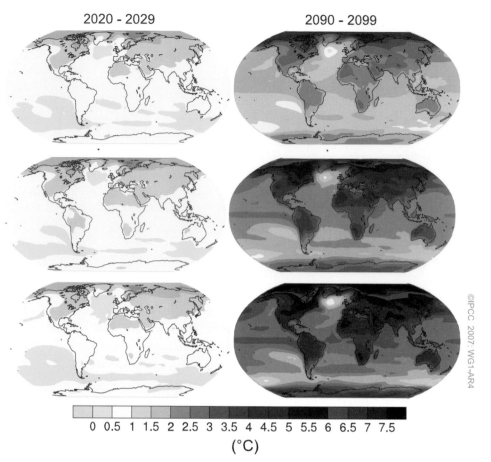

2020 - 2029   2090 - 2099

17.1 Projected surface temperature changes relative to the period 1980–99, showing the AOGCM multi-model average projections for the B1 (top), A1B (middle), and A2 (bottom) SRES scenarios averaged over the decades 2020–9 (left) and 2090–9 (right)
Graphic: IPCC, 2007 [i]

©IPCC 2007: WG1-AR4

0 0.5 1 1.5 2 2.5 3 3.5 4 4.5 5 5.5 6 6.5 7 7.5
(°C)

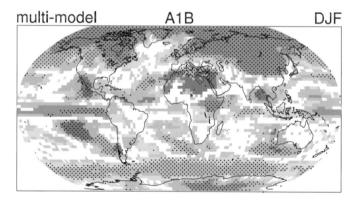

multi-model   A1B   DJF

17.2 Relative changes in precipitation (in percent) for the period 2090–9, relative to 1980–99. Values are multi-model averages based on the SRES A1B scenario for December to February (top) and June to August (bottom). White areas are where there is less than 66 percent agreement between models in the sign of the change, and stippled areas are where there is more than 90 percent agreement.
Graphic: IPCC, 2007 [ii]

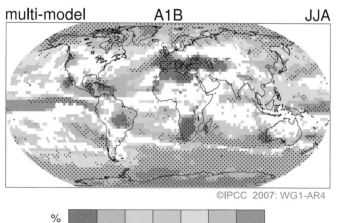

multi-model   A1B   JJA

©IPCC 2007: WG1-AR4

%   -20   -10   -5   5   10   20

and lifestyles around a climate free of even moderately high heat and that are unsuited to a future in which extreme temperatures and precipitation events become more frequent and intense. An example of the risks of failure to match a change in climate with a change in culture is the devastating 2003 European heat wave, to which about 50,000 deaths have been attributed (see Chapter 20).

Land Use, Increased Rainfall, and Stormwater Failure

More intense and frequent storms create a higher risk of overwhelming municipal stormwater systems. That risk is amplified when land-use decisions permit urban development on future floodplains or when deforestation and development upstream exacerbate runoff, leaving downstream areas vulnerable to floods. This brings knock-on risks of waterborne diseases, crop damage, and a loss in economic productivity.

Hurricane Katrina

In 2005, the Category 5 Hurricane Katrina tore through Louisiana and Mississippi in the southern United States. Its destruction revealed the weaknesses not just in the city's flood protection but also in the government and institutions in the city, leaving behind a legacy ranging from distrust to outright resistance. This is especially true for the poorer residents of the city, who arguably need more assistance in preparedness and recovery than their fellow citizens with greater economic and social capital. One writer observed that "Katrina exposed a broader social, political, and economic system that does not work for the poor."[2]

The failure of governments and other civic leaders to adequately anticipate and respond to one disaster can make the task of preparing for the next that much more difficult.

17.3 New Orleans skyline and areas flooded by Hurricane Katrina
Image: Caitlin Mirra/Shutterstock.com

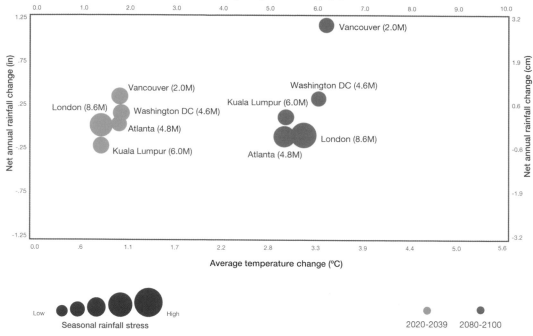

2020 - 2100 Changes
(Rainfall and Temperature)

17.4 Changes in rainfall and temperature anticipated for five sample cities and their populations (in millions)
Graphic: Arup, Data: authors' empirical analysis of World Bank Climate Change Portal data[iii]

Pests and Industry-Standard Construction

Climate change will expand the suitable ranges for many vector- and/or waterborne diseases and pests. Industry-standard construction can inadvertently exacerbate public health threats and damage to building materials (especially wood, fibers, and other natural materials) as local pest and disease profiles change.

Emergency Response Stress, Sector-Based Economic Decline

The double loss of climate change—from disaster recovery and costs of future resilience—illustrates why advanced planning is critical for communities, and even more so for communities already stretched by development needs. In addition, an extreme climate change event can expose the underlying tensions or dysfunction in infrastructure, economic, political, and social systems, making it all the more difficult to rebuild working relationships and civic trust in leaders to move forward after a disaster.

RECOGNIZE: ANTICIPATED CHANGES

The changes predicted in warmer and wetter climates generally include the following:

- Change in extremes (direct):
  - More intense storm events (wind strength, precipitation rate, etc.)
  - Flooding
  - Increased frequency, intensity, and duration of heat waves

- Change in averages (direct; see Figures 17.1 and 17.2):
  - Average and design temperatures (heat, cold, and humidity)
  - Design wind speeds
  - Average precipitation
  - Reduced diurnal/nocturnal temperature swings
- Increased average temperature and precipitation (indirect):
  - Reduced effectiveness of passive design
  - Increased cooling demand
  - Enhanced evaporation and drying of soils and wetlands
  - Increased erosion in surface waterways
  - Increased risk of land subsidence
  - Increased risk of vector- and/or waterborne disease
  - Increased risk of pest infestation
  - Economic downturn and social and political stress due to risk accumulation
  - For locations reliant on snowpack melt for water resources, less reliable water availability.

CHOOSE TO ACT: VULNERABILITY IN THE BUILT ENVIRONMENT

Figure 17.5 illustrates a selection of potential risks and coping capacities for the built environment in warmer and wetter regions. They have been plotted on a Vulnerability Map for an example community.

Example WARM AND WET Community with HIGH to VERY HIGH UN Human Development Index (HDI)

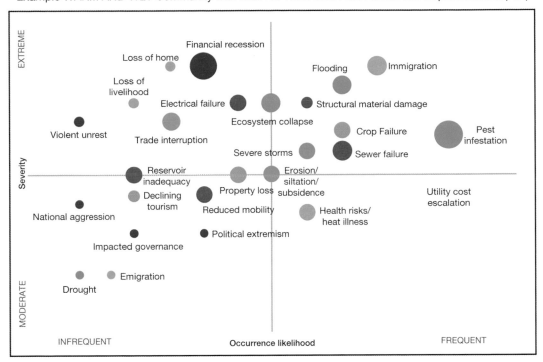

Example WARM AND WET Community with LOW to MEDIUM UN Human Development Index (HDI)

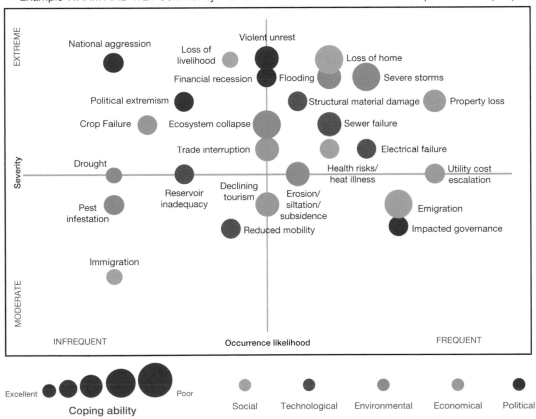

17.5 An example Vulnerability Map illustrating risk (the product of severity and likelihood) and coping capacity factors; the UN HDI has been used as a basis for coping ability—actual vulnerability will vary significantly within a community, and it is critical that assessments are of adequate rigor to account for such differences

Graphic: Arup

ACT SUCCESSFULLY: THE DESIGN, PLANNING, AND POLICY RESPONSE

Below is a summary of possible responses in warmer and wetter regions, organized according to the six-step approach and resilience framework described in Chapter 16. Those entries marked by an asterisk are common across one or more other regions and may therefore provide opportunities for broader interregional activity. Many locations likely to experience warmer and wetter future climates are also located in coastal areas (see Chapter 19).

1. Build Capacity (Immediate Action)

Before anything is built, invest in making better decisions and getting more useful information for pre- and post-adaptation needs.

- Educate and plan
  - *Social, political, institutional capacity** Understanding cultural systems is especially critical for developing appropriate resilience strategies that can be integrated with existing norms and find acceptance with formal and informal political, economic, and institutional stakeholders, especially in highly vulnerable urban poor communities.
  - *Flood mapping* Steps can be taken now, for example revising one-hundred-year and five-hundred-year floodplains to identify urban areas, transportation systems, energy and water infrastructure, and emergency response networks vulnerable to flooding.
  - *Climate adaptation plan** Prepare a draft plan that summarizes the risks, vulnerability, known information, and potential measures for adaptation in your region or regions. This can be a part of a broader climate change planning effort or a separate process focused solely on adaptation. See Chapter 16 for a sample outline.
  - *Climate adaptation training** Roll out programs to educate your community or organization on the implementation of the adaptation plan.
  - *Establish policy direction** Pilot, expand, and optimize policy to support a long-term process of improving adaptation.

2. Select the Right Site (Immediate Action)

Once the decision to build has been made, the location selected for development should be suitable.

- *Preserve natural hydrology** Preserve natural surface drainage systems and waterways, and preserve and/or enhance wetlands to promote "green" stormwater management and reduce flood risk.
- *Avoid development in high-risk areas** Recognize that areas that are relatively safe from flooding impacts today may be more exposed in the future:
  - Along coastal areas (see Chapter 19 for more detail)
  - Within the one-hundred- and five-hundred-year flood zones
  - In wetland areas or other low-lying marshy or riparian zones
  - In isolated areas with limited egress
  - At the base of a slope
  - At the top of an eroded or exposed slope
  - Areas exposed to damage from failure of adjacent structures (human-made and natural).
- *Watershed management* Promote watershed-wide partnerships to manage land-use change and development (especially of impervious areas) for stormwater management.
- *Site-appropriate development zoning and building codes** Resources available to policy makers include:
  - Design guidelines/standards
  - Comprehensive plan amendments
  - Area masterplan/plan amendments
  - Zoning overlays or code changes, such as the creation of a New Growth Boundary or moratorium in future flood areas
  - Risk-adaptation plan amendments
  - Policy/code language, legislative language.
- *Insurance availability** Insurance companies keep a close watch on the risks to development posed by climate change, including flooding, wind damage, and other storm-related damage. The availability and/or cost of insurance is likely to receive more attention in coastal communities as more extreme manifestations of climate change emerge (see Chapter 19).

Climate Change Resilience: Water Infrastructure in Wuhan

Wuhan is the capital of Hubei Province and a major Chinese metropolis, with an area of 8,400 square kilometers and a population of over eight million in 2008. The city is trisected by the third largest river in the world, the Yangtze, and its longest tributary, the Han River. Known as the "City with a Hundred Lakes," Wuhan's water bodies compose about a quarter of the city area. Wuhan is vulnerable to more frequent and severe climate change impacts, particularly heat waves, rainstorms, and floods.

With the Asian Development Bank and the city of Wuhan, Arup considered climate-related factors affecting existing and planned water infrastructure, both directly and indirectly, and strategies and

measures which Wuhan can use to adapt its water infrastructure to a changing climate. The information collected and analyzed empowers decision makers to properly adapt to climate change and address water resource management together.

The strategies and recommendations are intended to have a far wider applicability to other Chinese cities, with Wuhan serving as an initial pilot and model.

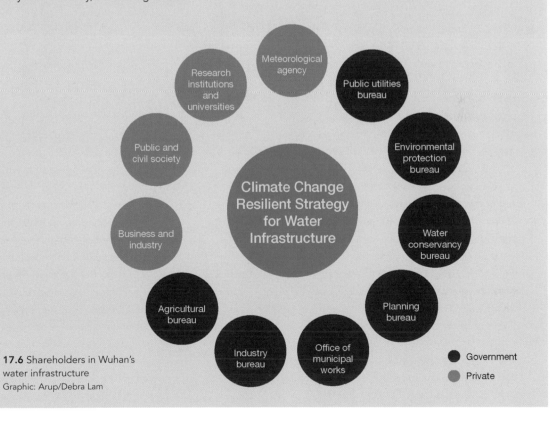

17.6 Shareholders in Wuhan's water infrastructure
Graphic: Arup/Debra Lam

201

When Is a Past Strategy No Longer the Best Option?

Just because one air-conditioning unit worked, adding thirty-five more isn't necessarily a better solution.

Misalignment of the response to a climate change impact can aggravate vulnerability and exacerbate the problem. In the instance of warmer temperatures, adding ever more air conditioners increases energy demand and these inefficient systems generate local heat, which adds to air-conditioning needs (and more energy demand), which creates more GHG emissions, which increase the magnitude of future climate change, which results in warmer temperatures. And so on.

17.7 Air-conditioning alleyway
Image: simonalvinge/Shutterstock.com

3. Build In Passive Survivability (Immediate Action)

After the project has been sited, pursue opportunities that don't require active interventions to be successful.

- *Survivability** Work toward the possibility of maintaining habitability with no or minimal resource input (energy, water, materials, etc.). Consider designing for a known point of systems failure, with easy access for inspection and repair to speed recovery.
- *Water: promote conservative design** Encourage conservative design of building drainage and site stormwater management.
- *Site and building design*
  - Control solar heat gain—reduce building cooling needs by reducing local microclimate heat islands, encouraging optimal building solar orientation, cool roofing, material selection, building shading, and facade and glazing shading.
  - Space conditioning—increased annual average temperatures and reduced diurnal/nocturnal swings can reduce effectiveness.
  - Occupant behavior—work to understand the needs and expectations of future building occupants and work with stakeholders to educate future occupants about passive design benefits and operations, enabling greater flexibility in acceptable conditions and design.
- *Regulations*
  - Stormwater management—legislators and community groups (like neighborhood associations) can enact site design mandates that require treating one hundred percent of projected stormwater runoff.

Village Homes: Resilience by Design

Reducing or eliminating additional site stormwater runoff reduces pressure on municipal systems. This is especially important where storm sewer and sanitary sewer infrastructures are combined and extreme rainfall events can cause flooding of sewer waste into waterways or other parts of the built environment. As discussed in Chapter 19, inadequate stormwater management can cause human health epidemics, a critical concern for communities in developing countries that do not have sufficient infrastructure, or where cultural practices (such as using open storm or sanitary sewers as refuse dumps) can exacerbate stormwater management issues.

When the 242-unit Village Homes "garden village" was built in Davis, California, in the 1970s *without* stormwater pipes in the ground, the developers were asked by the City to take out a bond in case their surface stormwater swales failed. When a fifty-year storm hit in the mid-1980s, Village Homes not only successfully managed its own stormwater, it also took overflow from the nearby conventional systems. Use of the surface drainage system saved nearly $1,000 per lot in infrastructure costs, resulted in minimal runoff to local rivers, and provides a network of green paths for the residents to this day.[3]

- Urban heat islands—reduce elevated air temperatures in urban areas by promoting urban green spaces, vegetation (green roofs, street trees), high-emissivity materials (materials that shed thermal heat gain, like wood, in contrast to black asphalt), and high-albedo materials (materials that reflect solar radiation, such as white roofs and light-colored concrete).

4. Design Active Resilient Systems (25+ years)
The built environment can't be entirely passive. Once the passive opportunities have been exploited, design in climate-appropriate, low resource intensity systems that can withstand partial or complete disruptions.

- *Consider future design conditions** Provide adequate space for phasing in mechanical systems or expanding cooling systems.
- *Create resilient infrastructure systems that can suffer failures without cascading failures ensuing** For example, multimodal transit systems that offer multiple paths of travel relieve the stresses caused by single points of failure due to flooding.
- *Employ stormwater best-management practices that slow down the rate of runoff and filter floodwater* Create vegetated green space between flood-prone areas and development that can expand and contract with runoff volumes. Avoid maladaptive measures such as impervious channeling that are likely to deflect high flows to other communities.
- *Create resilient building systems that can safe fail with reduced damage* Examples may include impervious ground-floor materials that are resistant to short-term floodwater exposure as well as pest damage.

5. Encourage Adaptable Buildings and Infrastructure (2+ to 200+ years)
Whatever decisions are made should be reasonably easy to change and adapt.

- *Design for how buildings learn*[4] By recognizing adaptability in layers, there is greater potential for building and infrastructure retrofits to be both lower cost and more effective. For example, policies may allow urban streets to be dotted with micro-parks that don't require new infrastructure but can lessen the local heat island effect.
- *Retrofit (i.e., teach) existing building stock* As our buildings age, capital reinvestment initiatives are an opportunity not only to beautify but to help the buildings better serve their occupants and owners. Examples include window retrofits, cool roofing, and air-cooled compressor-based air-conditioning.
- *Retrofit (i.e., teach) existing urban open space* To improve stormwater management and lessen urban heat islands, the street network should be gradually renewed with pervious surfaces, open-air pathways for circulation, cool materials, and shade structures. Opportunities for stormwater capture and reuse should be explored.

A Bluebelt of Green Infrastructure

The Staten Island Bluebelt, in the southernmost borough in New York City, is a stormwater management system for approximately one-third of Staten Island's land area. Since the early 1990s, the program has preserved natural drainage corridors, including streams, ponds, and other wetland areas, which allows them to perform their functions of conveying, storing, and filtering stormwater. The combined area of these sixteen watersheds totals approximately 10,000 acres. The City calculates that using natural systems in place of traditional sewers has saved taxpayers $80 million in infrastructure costs, raised property values, and restored damaged habitats, and it has plans to strategically expand the area networked to the system in coming years.[5]

17.8 A "bluebelt" of green infrastructure
Image: Pierdelune/Shutterstock.com

6. Manage Settlement and Retreat (50+ years)

While land-management practices, regional partnerships, local flood protection, and adapted building strategies continue to improve, some communities are already struggling with climate migrants. Residents are abandoning communities in search of better job opportunities or better housing elsewhere, and host communities have to absorb additional stress from immigrants. Simply put, there are two options for communities no longer able to cope with impacts:

- *Move buildings out of flood-prone areas* This can be a complete relocation to new land—which is costly not just in terms of money but also in terms of the identity that defines a community. It can also be in situ: in Galveston, Texas, after a 1900 hurricane devastated the city and took 8,000 lives, the most dramatic measure taken to prevent future storms from causing such devastation was to dredge sand to raise the city by as much as 17 feet (5.2 meters) above its previous elevation. The benefits of this effort were made clear in the aftermath of 2008's Hurricane Ike, in which elevated property was largely spared.[6]

- *Move infrastructure out of flood-prone areas* Moving infrastructure is a costly business. Energy infrastructure (power plants, substations, transmission and distribution networks), water infrastructure (potable water distribution networks, sanitary sewers, storm sewers, wastewater treatment networks), transportation networks (roads, rails, ports, airports), waste management (landfills, waste treatment facilities), telecommunication infrastructure (cables, cell towers, satellite dishes), emergency response infrastructure (government, medical, fire, police, civic centers)—urban areas provide a lot of services that can be interrupted, damaged, or destroyed by warmer and wetter climate change.

NOTES

1 "Environmental Refugees Unable to Return Home," January 3, 2010.

2 Sandra Crouse Quinn, "Hurricane Katrina: A Social and Public Health Disaster," *American Journal of Public Health*, 96(2), 2006: 204, http://www.ncbi.nlm.nih.gov/pmc/articles/PMC1470495/ (accessed June 11, 2012).

3 Judy Corbett and Michael Corbett, *Designing Sustainable Communities*, Washington, DC: Island Press, 2000.

4 Stewart Brand, *How Buildings Learn: What Happens After They're Built* (New York: Penguin Books, 1995).

5 "New York City Wetlands Strategy: Draft for Public Comment," New York: The City of New York, 2012, http://www.nyc.gov/html/planyc2030/downloads/pdf/wetlands_strategy.pdf (accessed June 12, 2012).

6 Amanda Ripley, "The 1900 Galveston Hurricane," *Time*, September 15, 2008, http://www.time.com/time/nation/article/0,8599,1841442,00.html (accessed June 12, 2012).

i S. Solomon, D. Qin, M. Manning, Z. Chen, M. Marquis, K. B. Averyt, M. Tignor, and H. L. Miller (eds.), *Climate Change 2007: The Physical Science Basis: Working Group I Contribution to the Fourth Assessment Report of the Intergovernmental Panel on Climate Change*, New York: Cambridge University Press, 2007, Figure SPM.6, p. 15.

ii S. Solomon, D. Qin, M. Manning, Z. Chen, M. Marquis, K. B. Averyt, M. Tignor, and H. L. Miller (eds.), *Climate Change 2007: The Physical Science Basis: Working Group I Contribution to the Fourth Assessment Report of the Intergovernmental Panel on Climate Change*, New York: Cambridge University Press, 2007, Figure SPM.7, p. 16.

iii At http://sdwebx.worldbank.org/climateportal/index.cfm.

The historic Hoover Dam on the Colorado River and Lake Mead, Arizona.
It is estimated that the lake will never again reach historic levels
Image: Mike Flippo/Shutterstock.com

18

GUIDELINE CHAPTER

Designing for Hotter and Drier Climates

Cole Roberts

I met a traveller from an antique land
Who said: Two vast and trunkless legs of stone
Stand in the desert. Near them, on the sand,
Half sunk, a shattered visage lies, whose frown,
And wrinkled lip, and sneer of cold command,
Tell that its sculptor well those passions read
Which yet survive, stamped on these lifeless things,
The hand that mocked them and the heart that fed.
And on the pedestal these words appear:
"My name is Ozymandias, king of kings:
Look on my works, ye Mighty, and despair!"
Nothing beside remains. Round the decay
Of that colossal wreck, boundless and bare
The lone and level sands stretch far away.

                Percy Bysshe Shelley, "Ozymandias"

What majesty is the desert storm coming on a plain. Seen from afar, mountainous forms of gray frame the flat and dimensionless blue sky. Winds rush through the brittle grasses. The air is energized. Like a train barreling through town, the hard pounding of raindrops and hail strikes charge the ground. With it comes the smell of water, vegetation, wetted earth, and life. And then it's gone. Left behind, dappled pools evaporate in the resurgent Sun. Saline soil may leach into the pools, and as the water evaporates, a taste of salt lingers. It as though the land has wept swollen tears. It is then that you can sense the mournful nature of deserts. Explosively alive, tenuous, and thirsting.

Deserts are among the most inspiring environments of raw beauty in all the world. They attract recreation, art, tourism, and agriculture. They are a retirement destination for many, a harrowing passage for others. They are both fragile and dangerous.

To those of us who choose to live in the hot and dry climates of the world, life is one of extremes. Temperatures may swing 60°F (33°C) in 24 hours. A year of poor precipitation can mean water rationing and the threat of ravaging wildfires. Consecutive years (or decades) of poor precipitation can mean disaster. If welcome rains do come, they can fall so hard and fast that minutes later and miles away, amid bright sun and cloudless skies, dry riverbeds can flash flood and carry lives away.

Climate change will have great impact on the hot and dry climates of the world, partly because of the direct changes in climate and partly because of the indirect threats to culture and economy.

Many deserts of the world appear to be entering periods of extreme, prolonged drought, and yet the people and plants living in these regions have grown complacent during periods of relative wetness. During this time, population has been expanding. Will cities such as Phoenix, Arizona, see dramatic outflows of residents as the number of days of extreme heat increase, water becomes scarcer, and gasoline costs rise, or will technology and adaptation be sufficient bulwarks to maintain livability?

This chapter explores the climate change predictions, risks, and design guidelines for successful adaptation in hot and dry climates. It is intended to be read in conjunction with Chapter 16 and also to function independently as a ready reference for those living or working in hot and dry regions. Although not all of this chapter will be immediately relevant to all hot and dry regions, the content serves as a starting point for forming a suitable adaptation response. As you read, consider how your region compares, imagine the people you know choosing (or failing) to act, and imagine the plausible futures that are even now being shaped.

RISK ACCUMULATION

Hot and dry climates are characterized by a unique combination of risks (i.e., sequential hazards or concatenated threats).

High Temperatures, Water Scarcity, and Nonresilient Urban Form

The default response to high temperatures in dry climates is to consume more water (for drinking, irrigating, and evaporative air-conditioning). As temperatures rise, the increased demand for water further strains already threatened resources. The intensity of this threat is exacerbated by the remarkably poor development of many cities for passive cooling and water efficiency. Many desert cities have been built whole or significantly expanded within the past fifty years, during a period when modern design and planning paid little attention to water efficiency, solar lot orientation, and thermally massive construction. The result is that the resilient form of old cities

Complacency along the Colorado

The next time you take a sip of coffee in the morning, imagine you're drinking a bit of the 5.4-million-year-old muddy Colorado River. Your single sip would consume roughly 250,000 years of the river's history. To sip the last 100 years, you would be rationed to less than a tenth of a drop. Not even enough to wet your tongue. Yet, it is this less-than-a-tenth-of-a-drop that forms the basis for southwestern U.S. water law and the 1922 Colorado River Compact—the agreement that apportions the river's water to its various dependents. When signed in 1922, the estimate of river water flow was 16.4 million acre-feet (maf) per year. Since then, flow has varied by a factor of five, from about 5 maf (1977) to 25 maf (1984). Tree ring studies have shown that average flow over the past three centuries has been 13.5 maf and varies widely from 4.4 maf to over 22 maf. Most recently, the 2000–4 drought was the most severe multiyear drought on record, with an average annual flow of 9.6 maf over those five years. Since actual depletions (use plus evaporation) are now about 14 maf annually, and given that some states' apportionments are still not fully consumed, the river's flow is already overapportioned and those who depend on it are likely faced with reductions in Colorado River supply of up to 75 percent during periods of extreme extended drought!

In 2007, a committee of scientists convened by the National Research Council published "Colorado River Basin Water Management: Evaluating and Adjusting to Hydroclimatic Variability."[1] In the report, the committee highlighted the historical "normality" of severe droughts in the Colorado River basin. It went on to predict more severe droughts than the tree ring record due to "temperature trends and projections," referring to progressively warmer future climate in the basin, which would lead to reduced snowpacks, earlier snowmelt, greater evapotranspiration, and lower streamflows. The effects of this warming will be superimposed on the natural variability described by the tree ring data, worsening the impact of "normal" drought. In effect, creating a new "normal" of variable flow and extreme drought.

Between 2000 and 2006, the seven states of the Colorado basin added five million people, a 10 percent population increase. Without changes in current management practices, there is as much as a 50 percent chance of fully depleting all of the Colorado River reservoir storage by 2050, according to a scientific study coauthored by NOAA-funded scientists. And 2050 is as near as the 1970s are fresh in our memory—just four decades separate us. The buildings and infrastructure planned and constructed today will certainly be around to see it, as will many of us.

18.1 The Central Arizona Project is designed to bring about 1.5 million acre-feet of Colorado River water per year to Pima, Pinal, and Maricopa Counties
Image: Tim Roberts Photography/Shutterstock.com

2020 - 2100 Changes
(Rainfall and Temperature)

18.2 Changes in rainfall and temperature anticipated for five sample cities and their populations (in millions)
Graphic: Arup,
Data: World Bank Climate Change Knowledge Portal

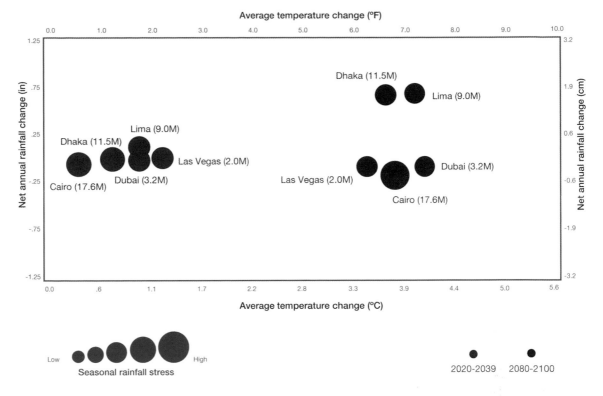

Seasonal rainfall stress — Low / High

2020-2039   2080-2100

is the exception rather than the rule. New city design has resulted in wide roads, dark (heat-absorbing) and impervious surfaces, and buildings poorly prepared for higher temperatures.

City Growth during the Age of the Automobile

The second combination of risks lies in the growth pattern of hot and dry cities—growth that has occurred during a period of low fuel costs and a cultural shift away from mass transit and dense urban form toward the automobile and sprawling development. As a result, hot and dry cities have a vulnerability to wildfires (due to proximity to combustible vegetation) and a uniquely high dependency on automobiles for transportation and access to necessities like food, health care, and employment. As fuel costs rise and public transit is hamstrung by low density, vulnerability will increase disproportionately among the elderly, young, pregnant, and disabled. Transportation will also take a greater proportion of the income of lower-income wage earners, including teachers, service staff, and the part-time employed.

Energy, Water, and/or Food Import Dependence

The third combination of risks originates from the dependency of hot and dry cities on imports of energy, water, and/or food for their economic life blood. A leading reason for the collapse of island civilizations of the past (including Greenland, Easter Island, and the Pitcairn Islands) was because they were cut off from their sources of trade. Similarly, desert cities often feel more like islands than oases. As energy, water, and food stress occur in trade partner cities, desert cities are expected to incur consequentially greater stresses.

RECOGNIZE: ANTICIPATED CHANGES

The changes predicted in hot and dry climates generally include the following:

- Change in extremes (direct):
  - Extreme heat
  - Extreme storms (flooding and dry lightning)
- Change in averages (direct):
  - Increased average temperature
  - Decreased average rainfall
- Indirect changes:
  - Rising commodity prices
  - Enhanced evaporation and drying of soils and wetlands
  - Economic downturn and social and political stress due to risk accumulation
  - For locations reliant on snowpack melt for water resources, water availability may become less reliable.

CHOOSE TO ACT:
VULNERABILITY IN THE BUILT ENVIRONMENT

Figure 18.3 (overleaf) illustrates a selection of potential risks and coping capacity for the built environment in hotter and drier regions. They have been plotted on a Vulnerability Map for an example community.

Example HOT AND DRY Community with HIGH to VERY HIGH UN Human Development Index (HDI)

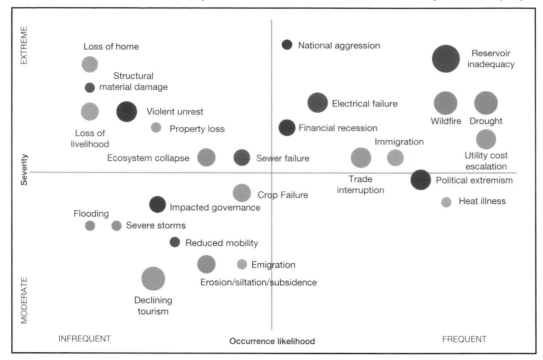

Example HOT AND DRY Community with LOW to MEDIUM UN Human Development Index (HDI)

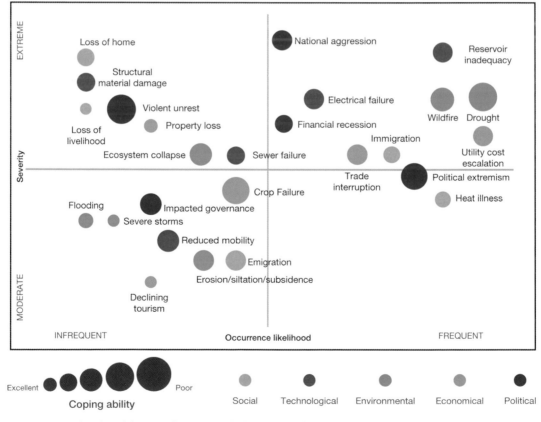

18.3 An example Vulnerability Map illustrating risk (the product of severity and likelihood) and coping capacity factors; the UN HDI has been used as a basis for coping ability—actual vulnerability will vary significantly within a community, and it is critical that assessments are of adequate rigor to account for such differences
Graphic: Arup

ACT SUCCESSFULLY:
THE DESIGN, PLANNING, AND POLICY RESPONSE

Below is a summary of the design and planning response for the built environment in hotter and drier regions. Those entries marked by an asterisk are common across one or more other regions and may therefore provide opportunities for broader interregional design integration.

1. Build Capacity (Immediate Action)

Before anything is built, invest in making better decisions and getting more useful information.

- Educate and plan
  - *Climate adaptation plan** Prepare a draft plan that summarizes the risks, vulnerability, known information, and potential measures for adaptation in your region or regions. This can be a part of a broader climate change planning effort, or a separate process focused solely on adaptation.
  - *Climate adaptation training** Roll out programs to educate your community or organization on the implementation of the adaptation plan.
  - *Establish policy direction** Pilot, expand, and optimize policy to support a long-term process of improving adaptation.
- Decision making
  - *Value water in life-cycle analysis* Only air can compete with water for a form so highly prized but little valued. To ensure good decision making in hot and dry climates, it is an imperative that system choices consider direct and indirect water consumption impacts. Life-cycle cost analysis should include the value of water consumed by each alternative. Where water costs are artificially depressed in value due to the sunk investment of large infrastructure projects (as in Arizona) or politically stabilizing subsidized pricing (as in Saudi Arabia), a reasoned estimate of the true value and marginal cost of securing new supplies should be considered for inclusion in the analysis.

2. Select the Right Site (Immediate Action)

Once the decision to build has been made, the location selected for development should be suitable.

- Flooding
  - *Plan for changes in stormwater intensity** Increased volumetric flow rates should be the basis of design as new and retrofit infrastructure is being built.
- Water availability
  - *Do not build in areas that have an unresolved water supply shortage greater than 50 percent within the lifetime of the development* California has a provision requiring developers of large projects (over five hundred residential units) to demonstrate that there will be an adequate water supply for twenty years before a building permit is issued. (Note this measure does challenge the majority of new nonreplacement development in the world's hot and dry climates.)
- Fire danger
  - *Do not build in areas susceptible to wildfire* (Note this measure does challenge the majority of suburban/rural development in the world's hot and dry climates, especially those in forested subregions.)

3. Build In Passive Survivability (Immediate Action)

After the project has been sited, pursue opportunities that don't require active interventions to be successful.

- Planting and water features
  - *Water efficient (xeric) plantings should be instituted into law* This is not the "zero"-scape of rock lawns and bright concrete patios, but an attractive native and adapted landscape that is drought tolerant and biologically active, and promotes local cooling through shade and evapotranspiration.
- Sun and shade
  - *Build in passive cooling** Incorporate the multitude of measures that allow buildings to compensate for the climatic extremes. These measures may include proper east–west orientation, reflective roofing, thermally massive materials, reduced glazing areas (especially on west faces), exterior window shades, shaded arcades, and high-performance glazing.
  - *Heat island reduction** Through urban planning and building design, temper the formation of heat islands, which artificially

Adapting to Change Down Under

If you're looking for a good example of adaptation planning for climate change, take some time to review the initiatives of the Australian government. The Australian government's position paper, "Adapting to Climate Change in Australia,"[2] sets out the government's vision for adapting to the impacts of climate change and proposes practical steps to realize that vision.

It outlines the Australian government's role in adaptation, which includes building community resilience and establishing the right conditions for people to adapt, taking climate change into account in the management of Commonwealth assets and programs, providing sound scientific information, and leading national reform.

The Australian government is supporting activities and financially investing to improve knowledge and build adaptation capacity. Highlights include:

- A $31 million Australian Climate Change Science Program
- A National Framework for Climate Change Science to set climate change research priorities
- $387 million for the Marine and Climate Super Science Initiative
- A $126 million Climate Change Adaptation Program helping Australians to better understand and manage climate change risks
- A National Climate Change Adaptation Research Facility
- $12.9 billion to secure Australia's water supply, in the single largest investment in climate change adaptation
- Support of Australian farmers as they adapt to climate change, through Australia's Farming Futures program
- The Caring for our Coasts policy, helping coastal communities prepare for and adapt to the impacts of climate change, including a national coastal risk assessment
- A Coasts and Climate Change Council established in late 2009 to engage with communities and stakeholders and to advise the government on key issues.

18.4 Australia is among the first governments to address the adaptation response to climate change at a national level
Image: Ian Woolcock/Shutterstock.com

raise the urban temperature. Shade hardscapes through building form and selective use of low water use trees. Select materials with a heightened reflectivity (albedo) and emissivity (ability to reemit captured heat). Focus heat island reduction efforts at the urban scale using vulnerability mapping, a process that highlights the locations of vulnerable populations.

- Rainwater and greywater
  - *Capture rainwater* Although not allowed in all regions, the local capture and use of rainwater in urban environments can be an excellent solution to both extend the available water for use and reduce the intensity of its flow in local stormwater networks.

4. Design Active Resilient Systems (25+ years)

Our built environment can't be entirely passive. Once the passive opportunities have been exploited, design in climate appropriate, low resource intensity systems that can withstand partial or complete disruptions.

- Building air-conditioning
  - *Consider compressor-based nonevaporative cooling* The energy efficiency advantage of evaporative systems when compared to air-cooled compressors is not as great as it once was. High efficiency compressor technology has advanced to close the gap.
  - *Select systems that have capacity to function in mixed modes (aka hybrid systems)** Such systems can more readily operate in a variety of modes, ranging from "off," to "on with natural ventilation," to "on with air-conditioned supply." They do not require the relatively cold air (55°F [13°C]) and water (44°F [7°C]) typical of most air-conditioning systems, thus dampening the severity of system disruptions due to compromised power or water supply. Examples include radiant cooling systems, underfloor air systems, and displacement air systems.
  - *Leverage low nighttime temperatures** The 30°F+ (17°C+) day-to-night temperature variation is uniquely appropriate for thermal energy storage strategies (ice, chilled water, or condenser water).

- *Wet/dry cooling towers* Where evaporative cooling towers are put to use, they should be hybrid towers capable of running the majority of the time in a dry "fluid cooler" mode and only occasionally in "wet" evaporative mode for extreme conditions. The water savings of these towers can exceed 95 percent of a conventional cooling tower.
- *Adopt future design conditions (corrected design day)** All systems should be specified in accordance with corrected climatic conditions appropriate to their system life (e.g., twenty-five years into the future for many heating, ventilation, and air-conditioning systems). The result may be slight increases to system size and shortened hours of natural ventilation modes during a typical year. Since many designers oversize systems already, a practice that can actually reduce performance, it is important that the system is right-sized for the future conditions, not simply further oversized.

- Building plumbing
  - *Design water-efficient systems* Flow rates on fixtures should be no greater than 1.5 gallons per minute.[3] Flush rates on toilets and urinals should be no greater than 1.28 and 0.125 gallons per flush, respectively. Hot water latency should be less than three seconds (achievable through a variety of measures), and where possible foot-operated faucets should be installed to free up hands during washing.
  - *Reuse greywater* To use water only once and send it down the drain is a profligate waste of opportunity. So unless there is a regional water balance initiative (as there is in Las Vegas, Nevada), which requires return water flows to the local water treatment plant, passive greywater redirection to local landscape can be a valuable strategy.
  - *Plan for local batch storage* Study cities like Delhi where water supply stress is acute and you'll see residences and office towers with on-site water storage. Given the limited hours that the municipal water supply service is turned on (in Delhi, one hour of flow twice a day), the small tanks (typically around five hundred gallons per household) serve to provide water for the hours when the municipal service is not flowing. They also provide a valuable emergency water source in the event of a disaster that may affect the municipal service for three or more days.

- Build to lessen wildfire danger
  - *Gutters* Eliminate gutters or design to minimize risk.
  - *Vented roofs* Avoid vented roofs or protect vents from ember entry.
  - *Specify low-combustible building materials* Class A roofing, with noncombustible decking and siding.
  - *Manage planting* Through distancing, plant selection, and other strategies lessen the impacts of planting on fire exposure.
- Create resilient infrastructure systems (create smart/resilient electrical and water networks that can suffer failures without ensuing cascading failures)
  - *Demand response** By shedding noncritical demands (electrical and water), the priority demands for life safety, economy, and environmental resilience can continue to be supplied in whole or in part.
  - *Diversify supply** Develop multiple supply sources (electrical and water), including the alternative supply "sources" of efficiency and closed-loop recovery, a failure in one supply can be compensated by another source.
  - *Informatics (making invisible information visible and visible information actionable)** By leveraging information technology and behavioral psychology, consumers and decision makers will have the opportunity to more effectively make decisions that enhance efficiency and cope with adversity.
- Plan resilient urban forms
  - *Establish, expand, and optimize diverse modes of transportation in concert with transit-oriented development** Diverse transportation modes supported by appropriate urban density create improved access to critical community services in the event of failures. They also support vulnerable populations, and encourage physical fitness. All of this can help minimize heat illness risks due to climate change while simultaneously improving general health, culture, and economic strength in our communities.
  - *Create urban vegetated space and community centers** These spaces provide alternatives for vulnerable populations facing heat illness threats. Air-conditioning in all climates and shade in dry climates in particular are among the best measures for reducing human mortality due to extreme heat.

5. Encourage Adaptable Buildings and Infrastructure (2+ to 200+ years)

Whatever decisions are made should be reasonably adaptable.

- *Design for how buildings learn*[4] By recognizing adaptability in layers, there is greater potential for building and infrastructure retrofits to be both lower cost and more effective. Where layers are crossed (e.g., embedded hydronic tubing in concrete structure), it is important that the lifespan of the layers is equal and/or that a retrofit strategy for the system with the shorter lifespan is anticipated.
- *Retrofit (i.e., teach) existing building stock** As our buildings age, capital reinvestment initiatives are an opportunity not only to beautify but to help the buildings better serve their occupants and owners. Examples include window retrofits, cool roofing, and air-cooled compressor-based air-conditioning.
- *Retrofit (i.e., teach) existing urban open space** To improve stormwater management and lessen urban heat islands, the street network should be gradually renewed with drought tolerant planting, cool materials, and pedestrian shade structures. Opportunities for stormwater capture and reuse should be explored.

6. Manage Settlement and Retreat (50+ years)

We only need to look back to the North American "dust bowl" of the 1930s to imagine what resettlement can look like. A false complacency stemming from a period of relative wetness had given rise to a belief that "rain follows the plow." When severe drought struck in 1933, the land blew away and economies collapsed. Within six years, 2.5 million Americans had relocated from nine U.S. states. On the other side of the globe and 60 years later, China has lost 2.6 million square kilometers to desertification since 1950 and faces accelerating desertification rates (currently at an additional 3,000 square kilometers per year). It is estimated that some 24,000 villages, 1,400 kilometers of rail lines, 30,000 kilometers of highways, and 50,000 kilometers of canals and waterways are subject to constant threat of desertification in China alone.[5]

Although land management and water use practices continue to improve, desertification and water availability are threats to limitless growth. Adaptation plans in hot and dry climates therefore need to make a reasonable effort to estimate their "right size" given available resources and scenario planning.

- *Managed settlement** As the carrying capacity of a region is approached, a variety of strategies may be used to manage growth while maintaining a strong economy. They include urban growth boundaries, development impact fees, and permit restrictions.
- *Passive retreat** As stresses increase, some resettlement will occur through market mechanisms. Commodity costs will increase. Water rights will be bought out by those willing to pay for them. This is likely to affect large water users (e.g., agriculture, golf courses, industrial users) most.
- *Active retreat** It is hard to judge if active retreat will be needed as in coastal regions. Presently, the authors assume that the question should continue to be posed even if the answer is yet unknown.

NOTES

1   Committee on the Scientific Bases of Colorado River Basin Water Management, National Research Council, Washington, DC: The National Academies Press, 2007.

2   Department of Climate Change, "Adapting to Climate Change in Australia: An Australian Government Position Paper," Canberra: Department of Climate Change, 2010; see also http://climatechange.gov.au/government/adapt.aspx.

3   Even at 1.5 gallons per minute, a typical shower will consume 15 gallons of water. In many modern global households (in e.g., Delhi, India or Kyoto, Japan), middle-class citizens wash with less than three gallons of water by using hot water containers and ladles. It helps if the bathing room is warm.

4   Stewart Brand, *How Buildings Learn: What Happens After They're Built*, New York: Penguin Books, 1995.

5   Dry lands occupy 43 percent of the world's land surface and are home to about 1 billion people. The science and case studies of desertification in North America, Australia, China, and Asia are presented in the United Nations report Y. Youlin, V. Squires, and L. Qi, *Global Alarm: Dust and Sandstorms from the World's Drylands*, Bonn: UNCCD, 2001.

Hong Kong is among the many cities at risk of rising seas and intense storms

Image: leungchopan/Shutterstock.com

19

GUIDELINE CHAPTER

Designing for Coastal Communities

Amy Leitch and Cole Roberts

The economic consequences of oceanfront catastrophes extend across the globe, burdening people who never admire the surf or cool off in an ocean breeze. More developed countries must set an example now for wise land use.

Editorial, *Engineering News-Record*, March 16, 2011[1]

Approximately three billion people—about half the world's population—live within 200 kilometers of a coastline. By 2025, that figure is likely to double.[2] Coastal communities are the engine of the world's society and economy: as birthplaces of our ancient cultural heritage and home to the most vibrant of the world's modern cultural centers, as centers of manufacturing and agricultural production, and as portals of world trade where goods and people exchange with the far reaches of the world.

Lloyd's of London and Risk Management Solutions predict that flood losses along tropical Atlantic coastlines would increase 80 percent by 2030 with about one foot of sea-level rise—in line with the conservative estimates of the 2007 report of the Intergovernmental Panel on Climate Change.[3]

LOCATION, LOCATION, LOCATION

The mantra of real estate agents the world over takes on new meaning when applied to climate change risk and resilience. Coastal communities are uniquely vulnerable to the effects of climate change, from physical damage due to sea-level rise and coastal flooding, to larger regional knock-on effects such as the breakdown in urban systems such as transportation, and energy and water supply. As coastal cities are pressed to accommodate current and future growth, available land begins to look more and more attractive. However, coastal development typically destroys natural systems that can help mitigate climate change impacts: reefs, barrier islands, beaches, sand dunes, and wetlands.

Isolated Small Island Developing States are especially vulnerable— if the sea level rises too high, people simply have nowhere left to go. Threats of more intense storms, accelerated erosion and the ongoing creep of sea-level rise may literally erase entire cultures from the map as their homeland is submerged. Low-lying island states are the proverbial canary in the mineshaft, as their geography makes them among the first witnesses to the effects of climate change. In addition, the resource constraints of islands make self-reliance in power and water supplies, health and food security more precarious, increase their vulnerability to knock-on effects, and could affect balances of power between communities or even nation states. Developing island nations that measure their carbon footprint in terms of livestock counts experience disproportionate effects of global climate change, and have fewer options for building resilience into their environment.[4]

As cities and urban areas grow, global assets are becoming increasingly concentrated geographically, escalating vulnerabilities to extreme climate-related events. The global trend towards population growth in coastal communities increases risks of loss—both from catastrophic events and from more frequent and intense lower-level events.

BETWEEN THE DEVIL AND THE DEEP BLUE SEA

Coastal communities have a vested interest in putting in place an early strategy to identify and respond to climate change to remain functioning places where people live, work, and play. Coastal tourism is an especially strong motivator as many coastal cities rely on the income earned from tourism in their communities. Between 1950 and 1995, the volume of global coastal tourism increased more than twenty times.[5] It has become one of the world's fastest growing industries. In the Seychelles, coastal tourism generates approximately 46 to 55 percent of national GDP,[6] while 85 percent of all U.S. domestic tourism revenue is attributed to coastal recreation.[7] But, unlike destination tourists who want to visit a specific place, coastal tourists have many options for where they choose to go. To remain competitive, coastal communities must ensure their amenities can weather climate change impacts.

Great Miami Hurricane

When Hurricane Andrew crashed ashore thirty miles south of Miami as a Category 5 hurricane in 1992, it caused a record $25.5 billion in damages (1992 dollars). However, accounting for coastal development and inflation, if the same storm hit today, analysts expect losses would double to over $50 billion (2005 dollars). One of the most devastating hurricanes in recent memory, Hurricane Andrew pales in comparison to the Great Miami Hurricane of 1926, which chartered a similar storm track almost sixty years earlier. The 1926 storm made landfall as a Category 4 storm and left behind about $760 million in damages (1926 dollars). If a similar storm struck today, in a more densely developed, crowded Miami, analysts estimate it would cause $130 billion of devastation (2005 dollars). And, if current trends continue—accounting only for additional property and inflation—a Great Miami Hurricane in 2020 would cause $500 billion in damages, a price tag that would leave the region, and the country, a devastating legacy for generations.[8]

19.1 1922 Miami River, Miami, Florida
Image: Fishbaugh, W. A. (William A.), Florida Photographic Collection

19.2 2010 Miami River, Miami, Florida
Image: fotog

The Maldives

The president of the Maldives is thinking about moving—his entire country.

The Maldives is an archipelago of almost 1,200 coral islands located south-southwest of India. As the lowest-lying nation in the world, most of its islands lie just 4.9 feet (1.5 meters) above sea level. With sea-level rise predictions in the Indian Ocean complicated by ocean currents, even a slight increase can aggravate tides and storm surges to devastate the nation.[9]

The Maldives is already working to reduce carbon emissions. The country announced plans to reach carbon neutrality by 2020, and in 2011 became the first country to crowdsource its renewable energy strategy from an open portal on the Internet.[10]

But as global emissions continue to rise, the government is also looking at the bleakest scenario: evacuation.

The tourist nation, with its white sand beaches and coral reefs, attracts wealthy tourists from around the world. With fears of inundation from rising sea levels, the country wants to set aside some of the $1 billion a year it receives from tourism to spend on buying a new homeland.

"We will invest in land," President Mohamed "Anni" Nasheed said. "We do not want to end up in refugee tents if the worst happens." The government reports it has broached the idea with several countries and found them to be "receptive."[11]

This represents a cautionary tale for other low-lying island nations, and raises a question for national sovereignty and global cultural heritage: can the Maldives remain "the Maldives" if the country relocates to another country?

President Nasheed was forced to step down on February 7, 2012, following a coup led by opposition figures. Although the coup is not directly linked to climate change risks, political instability is one potential impact of the destabilizing effects of climate change. It is unclear at the time of writing what implications this change in government may have for climate change resilience efforts.

19.3 Male, the capital of the Maldives
Image: Ho Yeow Hui/Shutterstock.com

	Timeframe	Cause	Predictability
Net extreme event hazard — Recurring extremes (Storm surge, storm tide)	Hours–days	Wave, winds, storm strength, coastal and offshore form	Moderate from observations, future very uncertain
Tide ranges	Daily–yearly	Gravitational cycles	Very predictable
Net regional mean sea-level rise (SLR) — Regional sea level variability	Seasonal–decadal	Wave climates	Not well know
Regional net land movements	Decades–millennia	Tectonics and compression	Predictable once measured
Regional eustatic SLR	Months–decades	Ocean warming, currents, climate	Existing effects observable, future effects uncertain
Global mean SLR	Decades–centuries	Climate change (temperature, ice melt)	Short-term extend current rates, future changes uncertain especially beyond 2100

19.4 Main contributors to net extreme event hazard and regional mean sea-level rise
Graphic: Australian Government Department of Climate Change, 2009

RECOGNIZE: ANTICIPATED CHANGES

Long-term and short-term contributions to the risk of coastal inundation can be categorized into changes in extreme events and changes in averages. The primary events are described in Figure 19.4.

The changes predicted in coastal communities generally include the following:

- Change in extremes (direct):
  - More intense storm events (wind strength, precipitation rate, etc.)
  - Flooding
  - More violent storm surges
- Change in averages (direct):
  - Sea-level rise—due to ocean currents, topography and tectonic movement, the increase is not uniform across the globe.[12]
    » Higher tides, higher storm surge, and increased saltwater intrusion into aquifers and estuaries, material degradation from increased saltwater exposure
- Average sea levels (indirect):
  - Compound impacts with other climate-related changes (precipitation, temperature, etc.)
  - Increased erosion in surface waterways
  - Increased risk of land subsidence
  - Increased risk of vector- and/or waterborne disease
  - Inland flooding due to reduced drainage capacity
  - Economic downturn and social and political stress due to risk accumulation.

RISK ACCUMULATION

Climate changes will have significant consequences for coastal environments, goods and services provided by coastal ecosystems, and coastal inhabitants.

Sea-Level Rise and Hydrologic Changes

Accelerated coastal erosion can damage ecosystems and wash away land, buildings, and infrastructure. Siltation of coastal waterways and potential changes in flows and currents can affect local hydrology and navigability of waterways, requiring more maintenance, intensive dredging, nourishment, and protection. Saltwater intrusion into fresh groundwater and surface waterways can poison ecosystems, affecting fresh water supply and productivity of agricultural land, and speeding material degradation in infrastructure and the built environment.

Ecosystem Squeeze, and Biodiversity and Livelihood Loss

Loss of coastal ecosystems and shifts in the distribution, abundance, and productivity of valuable marine habitats can affect fishing grounds and tourist destinations, with disastrous consequences for coastal economies, especially those dependent on natural resources. Engineered, hard protective infrastructure built to defend built assets can "squeeze" out coastal ecosystems by preventing natural migration. Left with fragmented, marginal habitat, coastal ecosystems become more vulnerable to stress and can suffer structural and biodiversity loss.[13]

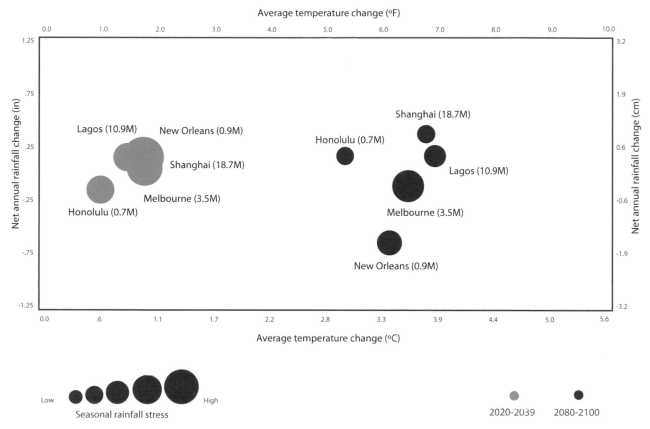

2020 - 2100 Changes
(Rainfall and Temperature)

19.5 Changes in rainfall and temperature anticipated for five sample cities and their populations (in millions)
Graphic: Arup, Data: authors' empirical analysis of World Bank Climate Change Portal data[ii]

Environmental Contaminants and Human Health

Other knock-on effects are less apparent, for example vulnerability to industrial legacies such as hazardous materials in contaminated soil. Many coastal communities have a history as polluted, industrial sites. As coastal communities experience increased frequency and higher levels of floods, contaminated land may become exposed to floodwaters, releasing buried toxins into surface or groundwater and exacerbating public health challenges.

Economic Dominoes and the Disaster Recovery Double Whammy

As in a domino chain, the effects of climate change are never isolated. As made devastatingly clear with the 2004 Indian Ocean tsunami, Hurricane Katrina in 2005, and the 2010 Japan tsunami, large-scale coastal environmental disasters can affect regional or even global economic and social systems. Knock-on impacts to buildings, transportation systems, energy and water supply, trade systems, food security, sanitation and human health can result in debilitating economic downturn. This double loss—of disaster recovery and costs of future resilience—illustrates why advanced planning is even more critical for coastal communities and the interconnected regional and global systems that rely on their productivity.

TABLE 19.1

Summary of the most important consequences of flooding in New York, Jakarta, and Rotterdam[14]

Graphic: Aerts et al., 2009

	New York	Jakarta	Rotterdam
Infrastructure/ transportation	Increased damage to and shutdown of rail and subway systems	Permanent inundation of low-lying areas and business district	Port facilities and railroads compromised
	Temporary inundation of low-lying business district (e.g., Wall Street) and transportation hubs	Decreased clearance under bridges	(Petro)chemical industry and increase of hazardous waste
	Decreased clearance under bridges	Structural damage due to coastal flooding	Decreased clearance under bridges
	Structural damage to infrastructure due to storm surge and wave setup	Temporary flooding of roads and airport	Structural damage to infrastructure due to storm surge and wave setup
	Increase in delays on public transportation and low-lying highways	Saltwater intrusion into freshwater resources. Increased damage to infrastructure not designed to withstand saltwater exposure	Delays in shipping due to closure of storm surge barrier
	Increased saltwater encroachment into local freshwater aquifers. Increased damage to infrastructure not designed to withstand saltwater exposure		
Waste/water infrastructure	Wastewater treatment plant, street, basement, subway, and sewer backups and flooding	Street, basement, and sewer flooding	Saltwater intrusion, loss of drinking water quality
	Increased pollution runoff from brownfields and waste storage facilities. Alternations to the flushing characteristics of the harbor and surrounding waterways	Increased pollution runoff	Diminished sewer system capacity
Nature/ environment	Intensification of the rate and extent of coastal erosion (damage to beach and salt marshes)	Beach erosion through degradation and coral reefs	Beach erosion
	Inundation of wetlands (loss of coastal wetlands and their associated ecological resources)	Increase in number of endangered species	
	Submergence and burial of shellfish beds by large volumes of sediment transport	Increase of frequency and intensity of flash floods and landslides	
		Increased rate of sedimentation	
Agriculture/ fisheries	Disruption of commercial and recreational fishing activities. Destruction of fishery/shellfishery habitats, migratory patterns	Polluted water may enter economically important fishponds	Saltwater intrusion may impact agricultural production, in particular the very salt-sensitive and very high-intensity production in greenhouses
	Damage to aquaculture facilities and loss of captive fish populations	Continued decrease of paddy production and decrease in maize production due to floods	

TABLE 19.2
Summary of adaptation measures to reduce impacts from coastal flooding due to climate change in New York, Jakarta, and Rotterdam[15]
Graphic: Aerts et al., 2009

Measure	New York	Jakarta	Rotterdam
Nature restoration/ augmentation	Beach nourishment, wetland and estuary restoration, salt marsh restoration	Improving the shoreline by mangrove conservation	Beach nourishment, water retention areas
Engineering protective measures	Sea walls, flood walls, groins, jetties, breakwaters, storm surge barriers	Levees, drainage systems, pumping stations, temporary dams	Levees, storm surge barriers
Freshwater storage	Staten Island Bluebelt Program, green roofs	Increasing greenery in the city, increasing capacity of existing watercourses through dredging and waste management	Green roofs, water plazas
Architecture	Flood-proofing basements and ground-level dwellings/buildings	Raised highways	Floating houses, adaptable waterfront development in old harbor area
Insurance	Available through NFIP (National Flood Insurance Program)	Available but market penetration is low	Not available

CHOOSE TO ACT:
VULNERABILITY IN THE BUILT ENVIRONMENT

Figure 19.6 (overleaf) illustrates a selection of potential risks and coping capacity for the built environment in coastal communities. They have been plotted on a Vulnerability Map for an example community.

ACT SUCCESSFULLY:
THE DESIGN, PLANNING, AND POLICY RESPONSE

In general, coastal communities have three primary options for managing climate change and building resilience. Discounting a fourth option, to do nothing, coastal communities have available in their planning and enforcement toolbox the ability to:

1. Protect the built environment by defending assets through soft and hard strategies
2. Accommodate climate change impacts in appropriate areas
3. Manage relocation of coastal assets, including people and infrastructure, to compensate for loss of ecosystem services.[16]

The [U.S.] National Institute of Building Sciences showed that every dollar spent on mitigation saves society about four dollars on recovery costs. Despite this evidence, nearly all U.S. coastal cities and towns lack adequate land use requirements and building code standards to realize these savings.

The Heinz Center and Ceres,
"Resilient Coasts: A Blueprint for Action"[17]

Example COASTAL Community with HIGH to VERY HIGH UN Human Development Index (HDI)

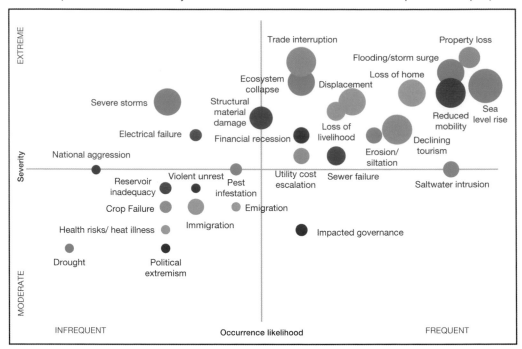

Example COASTAL Community with LOW to MEDIUM UN Human Development Index (HDI)

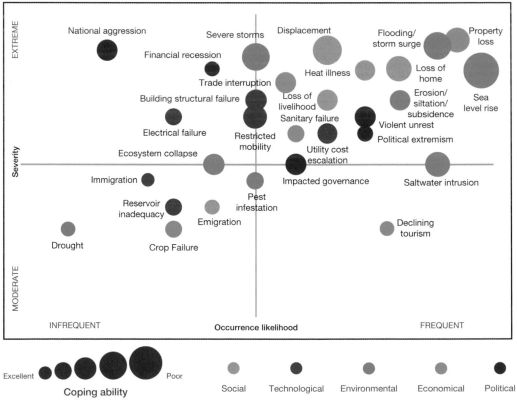

19.6 An example Vulnerability Map consisting of risk (the product of severity and likelihood) and coping capacity factors; the UN HDI has been used as a basis for coping ability—actual vulnerability will vary significantly within a community, and it is critical that assessments are of adequate rigor to account for such differences
Graphic: Arup

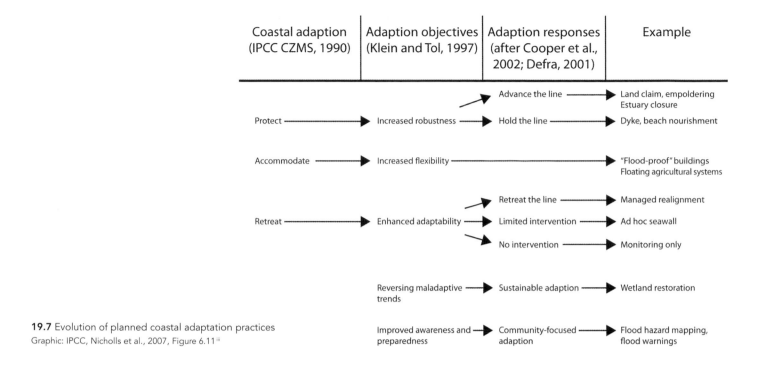

Coastal adaption (IPCC CZMS, 1990)	Adaption objectives (Klein and Tol, 1997)	Adaption responses (after Cooper et al., 2002; Defra, 2001)	Example
		Advance the line	Land claim, empoldering Estuary closure
Protect	Increased robustness	Hold the line	Dyke, beach nourishment
Accommodate	Increased flexibility		"Flood-proof" buildings Floating agricultural systems
		Retreat the line	Managed realignment
Retreat	Enhanced adaptability	Limited intervention	Ad hoc seawall
		No intervention	Monitoring only
	Reversing maladaptive trends	Sustainable adaption	Wetland restoration
	Improved awareness and preparedness	Community-focused adaption	Flood hazard mapping, flood warnings

19.7 Evolution of planned coastal adaptation practices
Graphic: IPCC, Nicholls et al., 2007, Figure 6.11 [iii]

Some regions, such as the southeast Florida coast, have already begun to explore how resilience efforts can be linked together across political boundaries for collective benefit.[18]

Below is a summary of possible responses in coastal communities, organized into the six-step approach and resilience framework described in Chapter 16. Those entries marked by an asterisk are common across one or more other regions and may therefore provide opportunities for broader interregional activity. Many coastal communities are also likely to experience warmer and wetter future climates, which are covered in Chapter 17.

1. Build Capacity (Immediate Action)

Before anything is built, invest in making better decisions and getting more useful information for pre- and post-adaptation needs.

- Educate and plan
  - *Social, political, institutional capacity** Understanding cultural systems is especially critical for developing appropriate resilience strategies that can be integrated with existing norms and can find acceptance with formal and informal political, economic, and institutional stakeholders, especially highly vulnerable urban poor communities. Working with long-time residents, people whose livelihoods are dependent on coastal natural resources, and community elders can highlight living systems that can be protected, restored, or enhanced to add to climate

Community Managed Sarstoon Temash Conservation Project

The Community Managed Sarstoon Temash Conservation Project (COMSTEC) in Belize is one example of a successful project to engage indigenous communities to create a comprehensive map of natural and sociocultural resources. Originated by five indigenous communities (four Q'eqchi' Maya groups and one Garifuna), the project has been supported by the International Fund for Agricultural Development and the World Bank since 2002, with the objective of identifying and preserving their ancestral lands. The project outcomes have supported the gathering of comprehensive baseline data on flora, fauna, soils and geology, hydrology, socioeconomic situation, and indigenous traditional knowledge. This data is also critical for understanding the characteristics of the local coastal ecosystems and social assets, especially in developing country communities, to help identify vulnerabilities to climate change impacts and develop feasible resilience strategies.[19]

change resilience. These efforts also bring the added bonus of empowering, training, and promoting active participation in resilience building among some of the more vulnerable community members.[20]

- *Monitoring* Monitoring changes in observed sea level, precipitation, and the frequency of extreme events can help prioritize investment, but cities and developers don't need to wait until the impacts are upon them.

- *Flood mapping* Steps can be taken now, such as revising one-hundred-year and five-hundred-year flood plains to identify urban areas, transportation systems, energy and water infrastructure, and emergency response networks vulnerable to flooding.

- *Climate adaptation plan** Prepare a draft plan that summarizes the risks, vulnerability, known information, and potential measures for adaptation in your region or regions. This can be a part of a broader climate change planning effort, or a separate process focused solely on adaptation. See Chapter 16 for a sample outline.

- *Climate adaptation training** Roll out programs to educate your community or organization on the implementation of the adaptation plan.

- *Establish policy direction** Pilot, expand, and optimize policy to support a long-term process of improving adaptation.

- *Knock-on human health impacts* In developing countries, natural disasters can make acute underlying public health challenges. Immediate threats, such as accidental electrocution risks associated with flooding where there is informal electricity transmission and distribution or less stringent or enforced protection, can hamper rescue efforts. Water-related diseases like dengue fever, cholera, leptospirosis, and diarrhea can create public health epidemics that overwhelm flood response efforts.

2. Select the Right Site (Immediate Action)

Once the decision to build has been made, the location selected for development should be suitable.

- *Preserve natural hydrology* Preserve natural surface drainage systems and waterways, and preserve and/or enhance wetlands to promote "green" stormwater management and reduce flood risk.

- *Avoid development in high-risk areas** Recognize that areas that are relatively safe from sea-level impacts today may be more exposed in the future. Critically, avoid development in these high-risk areas:
  - Along coastal areas
  - Within the one-hundred-year flood zone—and in the five-hundred-year flood zone if it can be avoided
  - In wetland areas or other low-lying marshy or riparian zones
  - In isolated areas with limited egress
  - At the base of a slope
  - At the top of an eroded or exposed slope
  - Areas exposed to damage from failure of adjacent structures (human-made and natural).

- *Watershed management* Promote watershed-wide partnerships to manage land-use change and development (especially of impervious areas) for stormwater management.

- *Site-appropriate development zoning and building codes** Resources available to policy makers include:
  - Design guidelines/standards
  - Comprehensive plan amendments
  - Area masterplan/plan amendments
  - Zoning overlays or code changes, such as the creation of a New Growth Boundary or moratorium in future flood areas
  - Risk adaptation plan amendments
  - Policy/code language, legislative language.

- *Insurance availability** Insurance companies keep a close watch on the risks to coastal development posed by climate change. The availability of insurance is a critical driver for economic growth and the ability of people, businesses, and governments to recover from loss in the event of damage. With climate change, insurance companies are being forced to review their exposure to vulnerable

Treasure Island

Treasure Island, located in the heart of San Francisco Bay, California, is a 535-acre man-made island built to host the 1939 World's Fair. Arup worked with SOM, premier developers and the City of San Francisco to create a masterplan for a transformation of the island into a sustainable new community. With a low elevation and exposure to the bay on all four sides, future sea-level rise was an important design consideration for the new development. But instead of building up defenses as the redevelopment is constructed, Arup worked with the site's design guidelines to integrate tipping points at which observed changes in sea level over time would trigger new phases of increasingly protective natural and built systems. This long-term approach uses phased code requirements to ensure appropriate coastal defenses are constructed when needed, avoiding reliance on predictions made today of future conditions.

19.8 Treasure Island redevelopment area, San Francisco Bay, California
Image: kropic1/Shutterstock.com

Quy Nhon

In 2008 the Rockefeller Foundation embarked on a major climate change initiative that concentrates on building the resilience of Asian cities to today's changing environment. Arup was selected to provide program management services for a component of this initiative focusing on catalyzing attention, funding, and action to build climate change resilience in Asian cities—the Asian Cities Climate Change Resilience Network (ACCCRN).

One of the cities partnering with the ACCCRN initiative is Quy Nhon, Vietnam. The capital city of Binh Dinh province, Quy Nhon is located just south of the Ha Thanh River. Primarily composed of low-lying coastal land, the city is bordered by the shore of Thi Nai lagoon, Lang Mai Bay and the South China Sea to the east, and the Day Truong Son mountain range to the west, exposing the city to coastal climate change impacts while constraining its ability to manage loss of productive land. While most administrative authority, planning decision making, and project management decisions are currently carried out by Binh Dinh province, the city is growing as an economic and political power, and will soon be able to more actively incorporate climate resilience activities and strategies into planning, development, and urban operations.

The city is frequently affected by a range of water-related impacts, which are likely to be exacerbated by climate change:

- All areas of the city are affected by flooding, particularly the peninsular and coastal areas, and along the banks of Thi Nai lagoon.
- Flash floods and river flooding that originate in the mountains on the western side of the city are frequent during the rainy season.
- The city often also experiences storm surges and sea flooding along the coastline, in addition to storm-related flooding, leading to inundation of portions of the city from two sides.
- Sea-level rise and a projected increase in the frequency and intensity of storms will exacerbate flooding hazards in the city and intensify saline intrusion and erosion issues.

To protect the city's drainage network, the city prohibits construction that negatively affects the size, water storage, and control capacity of water bodies or that is within 30 meters of the edge of water bodies. Where construction is pursued, planned elevations must be higher than the common severe flood levels. The province is also proactively working to address the potential changes to flood frequency or inundation depth expected with climate change, and is conducting experimental land-use planning with integration of climate change in two districts.[21]

areas. Five years after Hurricane Katrina devastated a large part of the Gulf Coast of the United States, homeowners in coastal areas across the eastern United States have seen the result of insurance companies' reevaluation of exposure, represented by an increase in insurance rates—or even an end to the ability to secure insurance at all. Home insurance rates for some coastal areas have shot up 30 percent or more.[22] In a sense, the availability and/or cost of insurance is an external check prompting developers to "select the right site" for projects, and is likely to receive more attention in coastal communities as more extreme manifestations of climate change emerge.

3. Build In Passive Survivability (Immediate Action)

After the project has been sited, pursue opportunities that don't require active interventions to be successful. Strategies can be phased in to offset high first costs.

- *Soft engineering* Natural systems require very little maintenance and have minimal repair costs compared to hard infrastructure. Through long-term protection, restoration, or enhancement of vegetated shoreline habitats, adaptation alternatives that favor natural approaches are generally recommended to stabilize the shoreline and reduce flooding risk. These "no regrets" strategies provide additional resilience for natural-resource dependent livelihoods even in the absence of climate change.[23]

- Living shorelines—Perhaps except in intensely developed areas, natural, or "living," shorelines are one of the most effective approaches to managing erosion and flooding. Strategies include wetland restoration and protection of natural dunes, reefs, mangroves, estuaries, sea grass beds, and barrier islands to reduce storm surge, coastal zone setback lines, and greenbelt development zones. In addition to coastal storm and flood protection, they also provide a range of ecosystem services such as water storage, groundwater recharge, retention of nutrients and sediment, pollutant filtration, fisheries, food security, and livelihoods.

- Green stormwater management—Protect or restore natural hydrology and water drainage networks. Strategies include bioswales, rain gardens, and flood and stormwater runoff pathway interruption.

4. Design Active Resilient Systems (25+ years)

The built environment can't be entirely passive. Once the passive opportunities have been exploited, design in climate-appropriate, low resource intensity systems that can withstand partial or complete disruptions.

- *Structural or hard engineering* Where natural systems are not appropriate, structural and mechanical systems can be designed to accommodate changing environmental conditions. Active systems are typically "hard" solutions that lack the flexibility of natural, "soft" solutions. A flood wall is only useful if the flood waters remain lower than its top. Integrating hard and soft solutions can build redundancies into coastal protection systems.

  - Active resilient surge protection—Protected, restored or constructed wetlands, barrier islands, and reefs; mechanical interventions; beach nourishment, artificial sand dunes, and dune rehabilitation; seawalls, sea dikes, storm surge barriers, and closure dams—local stormwater retention can avoid overwhelming municipal stormwater systems.

  - Monitoring—Emerging information and communication technology is opening up innovative opportunities to create

The Thames Barrier

The Thames Barrier protects about £80 billon ($125 trillion) worth of buildings and capital infrastructure in London. Some 1.25 million people live or work in the at-risk area.

Designed in the 1970s and operational in 1982, the Thames Barrier is the world's second-largest moveable flood barrier, after the Oosterscheldekering in the Netherlands. Built across a 520-meter (570-yard) stretch of the Thames, the barrier is located downstream from central London and is used to regulate exceptionally high water from tides and storm surges in the Thames estuary, to reduce flood risk in the city.

However, scientists are concerned that rapidly rising sea levels could significantly shorten the expected lifetime of the structure and expose London to devastating flood risk. By November 2011, the barrier had been closed 119 times to protect London from flooding. Of the closures, 78 were for tidal surge conditions and 41 were to prevent rainfall or fluvial flooding.[24] The rate of average sea-level rise has almost doubled from 1.8 to 3.1 millimeters since the barrier was being designed in the 1970s, when climate change was not seen as a risk.

Under the UK government's estimates of less than one meter sea-level rise, the Barrier will reach its maximum preferred closure rate of 70 times a year by around 2082. "The defenses we have at the moment allow for sea-level rise and the tidal levels we're expecting by 2030. That is still some time away. However, it takes time to research, design and build tidal defences, so we're already planning how we can manage increasing flood risk in the estuary," an Environment Agency spokesman has said.[25]

The Thames Barrier is a feat of engineering and has been providing flood protection for central London for thirty years. It also provides a valuable lesson on changeable parameters. As a hard, built engineering solution, it has little flexibility for dealing with the changing nature of climate change: it is as tall as it is. With a total construction cost of around £534 million (£1.3 billion at 2001 prices) and an additional £100 million for river defenses,[26] the initial price tag may be too high for many cities. But with foresight and adequate planning, phased-in retrofits can extend the barrier's lifetime and spread investment costs over a manageable time frame.[27]

resilient infrastructure systems. Smart technology such as wireless sensors can be used to create resilient electrical and water networks with monitors and built-in redundancies that can fail in the event of a disaster without triggering cascading failures. Integrating such protection systems for power plants, water and sewage plants, ports, and transportation systems is especially critical to creating resilient urban systems.

5. Encourage Adaptable Buildings and Infrastructure (2+ to 200+ years)

Whatever decisions are made should be reasonably easy to change and adapt.

- *Land use*
  - Consider reduced effectiveness of protective natural barriers (e.g., reefs, rocks, or other coastal formations) due to elevated sea levels.
  - Groundwater extraction may accelerate land subsidence and aggravate locally perceived sea-level rise and coastal flooding from tides and storms.
  - Position civil service infrastructure (police, medical, fire, community care) to protect it from flooding and sea-level changes.
  - Make use of regional collaboration within watersheds to promote stormwater management best practices and avoid activities that can aggravate erosion from upstream areas and clog downstream drainage.
- *Recovery*
  - Designing managed, controlled points of failure into building and infrastructure can release pressure on energy, water, and transport systems safely during an extreme event and speed recovery efforts, as engineers have more ready access and fewer check points to investigate.
- *Buildings*
  - Precipitation and flooding
    » Reducing wastewater production can result in significant benefits to municipal water and drainage systems for coastal communities and reduce pressures from flooding and sea-level rise.

» Flood doors installed at entryways can reduce flood risk, but are effective only if flood levels remain below their set height.
» Elevate mechanical systems to protect against flooding—Protect air intakes from salt spray to avoid corrosion. Reduce flood risk for buildings through strategies such as drain sizing, backflow devices, flood doors, foundation elevation, and elevation of mechanical systems.
» Buildings in flood-prone coastal areas should integrate stilts or other mechanisms to elevate space such as "soft" or flexible ground-level floors to accommodate flooding without significant damage.
» Structural design should allow for new facade requirements for easy retrofit. Air bricks installed in crawl spaces can allow for air movement to reduce mold during normal weather events but block vents in case of flooding.
» Consider material resilience to saltwater corrosion and increased humidity, and inspect exposed infrastructure more frequently. Elevated groundwater levels and saltwater intrusion into surface and groundwater can flood subbasements and corrode materials, affecting the longevity of infrastructure. Focus on reducing absorptive materials, and selecting more resilient materials.
  - Wind
    » Facades and cladding should be designed to withstand appropriate wind loads, speeds, directions, and pressures.
- *Infrastructure*
  - Energy
    » Power stations have high cooling requirements and tend to be sited near coasts or large rivers near the flood plain, exacerbating regional vulnerabilities to failure. Large-scale energy generation sites should consider vulnerability from site location and location of necessary equipment, to protect them from flooding and sea-level changes.
    » A move toward the greater decentralization of energy production will help lessen the risk of complete failure should systems fail. However, this will require additional attention to protect each generation unit and associated

transmission and distribution network from flooding, storm damage, sea-level rise, and other risks.

– Water

» Design to increase conveyance capacity for sanitary and storm sewers, and separate combined systems. Risks of human and environmental exposure to contaminated flood water are a serious concern, raising public health threats.

» Design to protect from inland flooding as well as coastal flooding. Inland flooding can be especially acute in communities whose cultural traditions regard waterways as waste removers, or developing countries that have fewer restrictions or enforcement capacity to prevent waste material from clogging rivers and drainage infrastructure.

» Coastal areas in regions that are likely to experience reduced precipitation may augment potable water supplies with desalinization, but this is energy intensive, can disrupt saline levels in adjacent waters, and is located in vulnerable coastal areas.

» Long-term protection projects such as levees and barriers built across inlets, channels, ports, etc. can help protect inland areas. However, these hard infrastructure solutions are not fail-safe, and are beyond the budget of many cities in economically stressed areas. Incorporating redundancies can stagger damage, allow more time for evacuation and shoring up of defenses if the first line of defense fails, and lessen flooding throughout the entirety of the vulnerable area.

– Transport

» Longer-term masterplanning should consider alternative routes and extra capacity in addition to redundancy in case of emergency.

» Below-grade transport networks like tunnels and subways are especially vulnerable to flooding. Consider phased protection and increased pump capacity.

» Consider re-routing or abandoning key rail and highway or road connections highly vulnerable to coastal flooding.

» Increase dredging of ports and navigation channels to maintain or enhance navigability where water levels are predicted to drop (e.g., the Great Lakes and the Mississippi River), as well as in coastal areas with increased drainage from inland watersheds.

» Where sea-level rise reduces bridge clearance and affects air draft, increase the frequency of bridge openings, raise the clearance of new bridges or incorporate climate change effects into planning, and replace or relocate existing bridges that are no longer adequate.

6. Manage Settlement and Retreat (50+ years)

Moving development out of harm's way in a planned and controlled manner is a last-resort option for many coastal communities. Techniques such as land abandonment, infrastructure relocation, and development avoidance are intergenerational mechanisms that prepare for coastal effects of climate change. Although managed retreat may be more contentious politically and socially than accommodation or protection, it may be ecologically sustainable, allowing for ecosystem processes and retreat, and, at manageable scales, more financially sustainable as it avoids costs associated with protection, damages, and interruption to urban systems.

NOTES

1 "Lessons from Japan's Great Quake and Tsunami Show the Limits of Control," http://enr.construction.com/opinions/editorials/2011/0316-JapanQuakeTsunami.asp (accessed June 12, 2012).

2 Liz Creel, "Ripple Effects: Population and Coastal Regions," policy brief, Washington, DC: Population Reference Bureau, 2003, http://www.prb.org/Publications/PolicyBriefs/RippleEffectsPopulationandCoastalRegions.aspx (accessed June 12, 2012).

3 "Resilient Coasts: A Blueprint for Action," Washington, DC and Boston, MA: The Heinz Center and Ceres, 2009, http://www.heinzctr.org/Major_Reports_files/Resilient%20Coasts%20Blueprint%20for%20Action.pdf (accessed June 12, 2012).

4 National Climate Change Country Team, "Government of Samoa: First National Communication to the UNFCCC," Apia, Samoa: Department of Lands, Surveys and Environment, 1999, http://unfccc.int/resource/docs/natc/samnc1.pdf (accessed June 12, 2012).

5 Anne Platt McGinn, "Safeguarding the Health of the Oceans," Worldwatch Paper 145, Worldwatch Institute, 1999, http://www.worldwatch.org/system/files/EWP145.pdf (accessed June 20, 2012).

6 United Nations Environment Program, "Western Indian Ocean Islands," Africa Environment Outlook, United Nations Publications, 2003, http://www.unep.org/dewa/africa/publications/aeo-1/109.htm (accessed June 20, 2012).

7 James R. Houston, "The Economic Value of Beaches: A 2008 Update," Shore & Beach, 76(3), 2008: 23, http://beachlobbyist.com/files/Economic_Value_of_Beaches_(2008).pdf (accessed June 20, 2012).

8 Kenneth Chang, "In Study, a History Lesson on the Costs of Hurricanes," New York Times, December 11, 2005, http://query.nytimes.com/gst/fullpage.html?res=9F05E1D81131F932A25751C1A9639C8B63 (accessed June 12, 2012).

9 CNN, November 11, 2008, http://edition.cnn.com/2008/WORLD/asiapcf/11/11/maldives.president/index.html (accessed June 12, 2012).

10 Duncan Clark, "Maldives Crowdsources 2020 Carbon Neutral Plan," The Guardian, September 22, 2011, http://www.guardian.co.uk/environment/2011/sep/22/maldives-help-carbon-neutrality-plan (accessed June 12, 2012).

11 CNN, "Sinking Island's Nationals Seek New Home."

12 NASA, "PIA11002: Portrait of a Warming Ocean and Rising Sea Levels: Trend of Sea Level Change 1993–2008," California Institute of Technology Jet Propulsion Laboratory, http://photojournal.jpl.nasa.gov/catalog/PIA11002 (accessed June 20, 2012).

13 U.S. Agency for International Development (USAID), "Adapting to Coastal Climate Change: A Guidebook for Development Planners," 2009, p. 50, Table 3.2, http://pdf.usaid.gov/pdf_docs/PNADO614.pdf (accessed June 20, 2012).

14 J. Aerts, D. C. Major, M. J. Bowman, P. Dircke, and M. A. Marfai, "Connecting Delta Cities: Coastal Cities, Flood Risk Management and Adaptation to Climate Change," Amsterdam: VU University Press, 2009, p. 44–45, Table 5.1, http://www.rotterdamclimateinitiative.nl/documents/CDC/ConnectingDeltaCities.pdf (accessed June 20, 2012).

15 See note 14.

16 R. J. Nicholls, P. P. Wong, V. R. Burkett, J. O. Codignotto, J. E. Hay, R. F. McLean, S. Ragoonaden, and C. D. Woodroffe, 2007: "Coastal Systems and Low-lying Areas," in M. L. Parry, O. F. Canziani, J. P. Palutikof, P. J. van der Linden, and C. E. Hanson (eds.), *Climate Change 2007: Impacts, Adaptation and Vulnerability. Contribution of Working Group II to the Fourth Assessment Report of the Intergovernmental Panel on Climate Change*, Cambridge, UK: Cambridge University Press, 315–56, http://www.ipcc.ch/pdf/assessment-report/ar4/wg2/ar4-wg2-chapter6.pdf (accessed June 13, 2012).

17 See note 3 above.

18 See http://www.southeastfloridaclimatecompact.org/.

19 C. Sobrevila, "The Role of Indigenous Peoples in Biodiversity Conservation: The Natural but Often Forgotten Partners," Washington, DC: World Bank, 2008, http://siteresources.worldbank.org/INTBIODIVERSITY/Resources/RoleofIndigenousPeoplesinBiodiversityConservation.pdf (accessed June 13, 2012).

20 See note 17 above.

21 ISET (eds.), "Asian Cities Climate Change Resilience Network: Responding to the Urban Climate Challenge," report, Boulder, CO: ISET, 2009, pp. 19–20, http://www.rockefellerfoundation.org/uploads/files/2d4557bc-6836-4ece-a6f4-fa3eda1f6c0c-acccrn_cop15.pdf (accessed June 13, 2012).

22 S. Block, "5 Years after Katrina, Homeowners Insurance Costs More," *USA Today*, August 26, 2010, http://www.usatoday.com/money/industries/insurance/2010-08-26-katrina26_CV_N.htm (accessed June 13, 2012).

23 http://www.usaid.gov/our_work/cross-cutting_programs/water/docs/coastal_adaptation/adapting_to_coastal_climate_change.pdf.

24 http://www.environment-agency.gov.uk/homeandleisure/floods/38359.aspx.

25 S. Connor, "Sea Levels Rising Too Fast for Thames Barrier," *The Independent*, March 22, 2008, http://www.independent.co.uk/environment/climate-change/sea-levels-rising-too-fast-for-thames-barrier-799303.html (accessed June 13, 2012).

26 See http://www.environment-agency.gov.uk/homeandleisure/floods/38353.aspx.

27 http://www.eastlondonlines.co.uk/ell_wp/wp-content/uploads/2010/03/thamesbarrier.jpg.

i   "Climate Change Risks to Australia's Coast: A First Pass National Assessment," Commonwealth of Australia, 2009, p. 23, Figure 2.2, http://www.climatechange.gov.au/~/media/publications/coastline/cc-risks-full-report.pdf (accessed June 20, 2012).

ii At http://sdwebx.worldbank.org/climateportal/index.cfm.

iii See note i in Chapter 11.

Rust infestation—the disease is anticipated to expand in
geographic impact as the climate warms
Image: Damian Herde/Shutterstock.com

20

GUIDELINE CHAPTER

Designing for Inland Communities

Afaan Naqvi and Cole Roberts

Disconnecting from change does not recapture the past. It loses the future.

Kathleen Norris,
poet and essayist

It is our future in which we will find our greatness.

Pierre Trudeau (1919–2000),
fifteenth prime minister of Canada

Whether one has traveled across the great plains of North America, the grassy steppes of Asia, the African savanna, or the rolling hills of Europe, or heard accounts from those who have, the picture is deceptively simple to paint but often hard to fully experience. The sense of openness and vastness is inspiring, regardless of the traveler or the listener. The austere beauty of flat fields and verdant rolling hillsides seems endless in all directions. No coastline is visible, horizon to horizon, and the land is often watered by irrigation, making threats such as sea-level rise and drought seem distant and unconnected. What, then, should those of us living in these communities care of climate change?

Perhaps the most broadly distributed of communities discussed in this book, inland communities are often considered the heartland of our nations—agricultural and pastoral centers of gravity, the food baskets of the world. Geographically, we build them on every continent and at all the latitudes, from equatorial to subarctic. Climatically, they are distant from the dampening effects of the seas and therefore experience great variations in rainfall and temperature, with cold winters and hot summers.

The distance from the seas also means the influence of the major ports is diluted in our inland communities, historically insulated to a greater degree than coastal regions from repeated waves of immigrants and cultural disruption. As a result, our inland communities often tend toward greater commonality of culture, religion, and politics. Although they tend to be more internally homogeneous, they are also more globally dissimilar from one another, whereas coastal communities are more locally mixed and heterogeneous, yet globally more akin to one another.

Those of us who live in inland communities tend to experience lower population densities than coastal regions, where economy and experience tempt youth, and although inland communities tend to have larger family sizes due to labor needs and religious doctrine, they are also often net losers of population, as our nations increasingly urbanize and become globally interconnected. Where cities do exist, they tend to sit alongside or at the confluence of the world's great rivers, which both flood and irrigate, setting up a unique relationship of hazard and dependency. Although these waters flow to the sea, the lack of visual and cultural connection to the world's oceans means that sea-level rise is less locally relevant.

This chapter explores some of the more prominent characteristics of these areas and summarizes the commonalities in adaptation response. It is intended to be read in conjunction with Chapter 16 and also to function independently as a ready reference for those living or working in inland regions. Since inland regions may be getting warmer and wetter or hotter and dryer, readers may also value reading the parallel chapters (Chapters 17 and 18, respectively).

Although not all of this chapter will be immediately relevant to all inland regions, the content serves as a starting point for forming a suitable adaptation response. As you read, consider how your region compares, imagine the people you know choosing (or failing) to act, and imagine the plausible futures that are even now being shaped.

RISK ACCUMULATION

Heat Index, Cultural Norms, and Nonadapted Buildings and Infrastructure

Rising temperature and humidity will have a significant impact on communities in inland regions that have developed their infrastructure and lifestyles around a climate free of even moderate heat. An example is the estimated 50,000 heat-related deaths experienced in the 2003 European heat wave, which are often incorrectly associated with extreme heat.[1] Rather, the majority of these deaths can be traced to above average but nonextreme temperatures coupled with a cultural and historical lack of air-conditioning in some regions, choice of construction materials (such as stone and concrete suitable for the historically cool climate), and simple timing (much of the community tends to vacation in the summer, leaving a vast number

2020 - 2100 Changes
(Rainfall and Temperature)

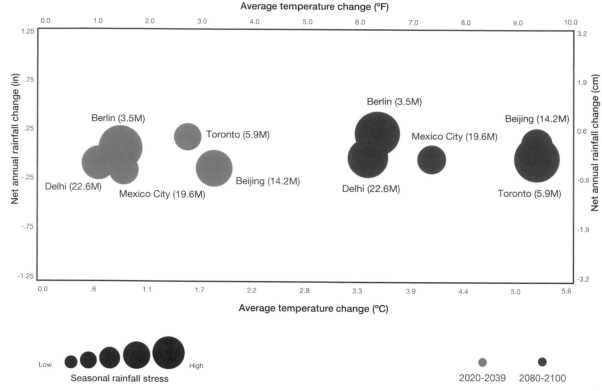

20.1 Changes in rainfall and temperature anticipated for five sample cities and their populations (in millions)
Graphic: Arup, Data: World Bank Climate Change Portal

of the elderly population unattended for long periods and a lack of adequate physicians locally). Since power disruptions are likely to coincide with higher heat levels (due to air-conditioning loads), the risk may accumulate uniquely when there are failures in electricity distribution.

Changing Patterns of Precipitation, Proximity to Rivers, and Sewer Infrastructure

Areas seeing more intense and frequent storms are at a higher risk of overwhelmed municipal wastewater systems. This brings not only the direct risk of local flooding, but the knock-on effects of waterborne diseases, crop damage, and a loss in economic productivity. In specific regions where development is at the water's edge, the risk is greater and extends to property damage and subsequent loss of value, and, in some unprepared communities, loss of life.

Low Density, Sprawl, and Automobile Dependency

The growth pattern of many inland regions mirrors that of hot and dry cities, in that growth happened during a period of low fuel costs and a cultural shift away from mass transit and dense urban form. The resulting automobile dependency and sprawl patterns result in an increased wildfire threat. The low density and more prominent rural life in these regions also creates greater distances between people, jobs, services, and schools, and the associated reliance on

and use of private transportation infrastructure and energy tends to be less efficient on a per capita basis. These communities will therefore also be less able to cope with a steep increase in energy generation and/or import costs while public transit is hamstrung by low density. As a result, vulnerability will fall disproportionately on the elderly, young, pregnant, and impoverished.

Monocrops, Wooden Structures, and Retreating Freeze Lines

With warmer winters, the encroachment of pests and diseases on cropland and wooden buildings is likely to grow, affecting one of the primary economic activities and the homes of many inland communities. This becomes an indirect threat to other communities that rely on inland regions for food import.

RECOGNIZE: ANTICIPATED CHANGES

The changes predicted in inland climates generally include the following:[2]

- Change in extremes (direct):
  - Extreme heat
  - Extreme storms, notably tornados and flooding
- Change in averages (direct):
  - Increased average temperature

– Increased or decreased average rainfall depending on specific region
- Indirect changes:
  – Heat fuel poverty
  – Transportation fuel poverty
  – Increased risk of building pest infestation and crop disease
  – Economic downturn and social and political stress due to risk accumulation.

CHOOSE TO ACT: VULNERABILITY IN THE BUILT ENVIRONMENT

Figure 20.2 (overleaf) illustrates a selection of potential risks and coping capacities for the built environment in inland communities. They have been plotted on a Vulnerability Map for an example community.

Urban or Rural

In many parts of the world, inland communities are dominated by smaller cities, towns, and often rural settlements. This can either heighten or alleviate climate change risks and should be considered in adaptation efforts. Communities in these areas may be distrustful of top-down programs and government intervention, making openness to externally sponsored adaptation efforts challenging. However, the resiliency of such communities can be high as members self-support each other and often have strong local institutions of leadership in times of crisis and threat.

ACT SUCCESSFULLY: THE DESIGN, PLANNING, AND POLICY RESPONSE

1. Build Capacity (Immediate Action)

Before anything is built, invest in making better decisions and getting more useful information.

- Educate and plan
  – *Climate adaptation plan* Prepare a draft plan that summarizes the risks, vulnerability, known information, and potential measures for adaptation in your region or regions. This can be a part of a broader climate change planning effort or a separate process focused solely on adaptation.
  – *Climate adaptation training* Roll out programs to educate your community or organization on the implementation of the adaptation plan.
  – *Establish policy direction* Pilot, expand, and optimize policy to support a long-term process of improving adaptation.
- Decision making
  – *Value carbon in life-cycle analysis* Energy in the United States was once abundant, inexpensive, and not factored into decisions made about the built environment. Since then, mass scientific, public, and political opinion has gone through phases of denial, recognition, and action, to the extent that building with energy efficiency in mind today is expected and often legislated, rather than scorned or just appreciated. Barring outliers, the denial and recognition phases associated with the carbon problem are largely over, especially in the scientific and political eye. However, only in the case of early voluntary movers and legislated or financially rewarded environments is the action phase implemented. In addition to helping mitigate future emissions, valuing carbon in life-cycle analysis will promote density, increase public transportation feasibility and reliance, and spur local market and capability growth, making inland communities more self-sufficient. Thereby, it will act as a vehicle for pre-adaptation and also result in less need for harsh changes later.

Example INLAND Community with HIGH to VERY HIGH UN Human Development Index (HDI)

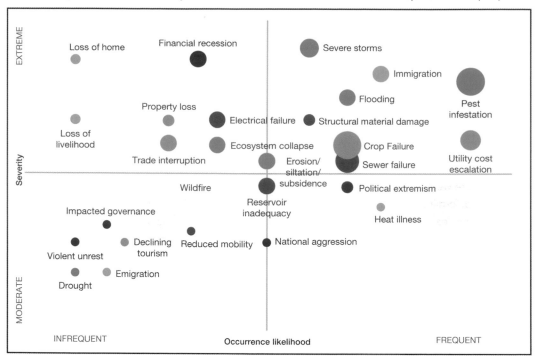

Example INLAND Community with LOW to MEDIUM UN Human Development Index (HDI)

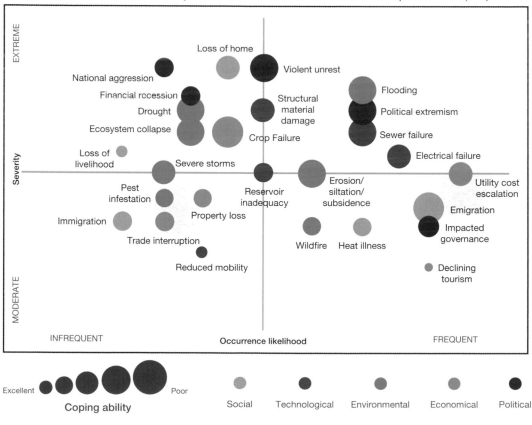

20.2 An example Vulnerability Map consisting of risk (the product of severity and likelihood) and coping capacity factors for inland communities; the UN HDI has been used as a basis for coping ability—actual vulnerability will vary significantly within a community, and it is critical that assessments are of adequate rigor to account for such differences
Graphic: Arup

Falls Lake, North Carolina

The worst drought witnessed in North Carolina's recorded history occurred in 2007 and 2008. It began in February 2007, and by December that year every county in the state was in drought. Normal water conditions were not restored until more than two years later, in December 2009.

During this period, there were several adaptation responses and activities that occurred, forming a recent case study for inland communities to consider and learn from. Starting with mandatory adaptation measures, the entire state adopted water restrictions and state agency water use was limited to essential purposes only. Next, the governor called on residents to curtail their water use by as much as 50 percent. Thirty cities and towns came close to running out of water, and in Siler City water was actually trucked in as a response to severely low water supply.

In the state capital, Raleigh, residents viewed the continued allowance of building permits and further development while they were being asked to cut their water use as irresponsible governance. Though some state council members did float the idea of a temporary pause in building and new construction activity, it was quickly met with resistance from other city officials, home builders' associations, and other local businesses. The absence of a city, state, or regional adaptation plan became more and more evident, as the drought continued and the interests of various stakeholders were expressed, and were rarely in alignment.

20.3 Falls Lake, North Carolina, during the 2007–8 drought
Image: © Bob Sowa

AFAAN NAQVI AND COLE ROBERTS

2. Select the Right Site (Immediate Action)

Once the decision to build has been made, the location selected for development should be suitable.

- *Plan for changes in stormwater intensity* Increased volumetric flow rates should be the basis of design as new and retrofit infrastructure is being built.
- *Fire danger* Do not build in areas that are or will become susceptible to wildfire within the life of the building.
- *Promote density and mixed use* Placing people close to jobs, services, schools, and other necessities will alleviate multiple stressors, including fire danger, energy and fuel price rises, and import dependency. It will also make central systems such as air-conditioning, waste collection and treatment, and energy recovery more feasible, further alleviating the adaptation stress caused by sprawl.

Density and Heat Islands

This chapter repeatedly makes the case for density, mixed use, and local or self-sufficiency. The authors recognize that, while these strategies alleviate climate change impacts and reduce the magnitude of adaptation needed later, they often seem to bring their own challenges. One such challenge is the heat island effect, which is known to raise temperatures in large cities by up to 5°F[3] and increase levels of illness in cities that already experience unusually high summertime temperatures.[4]

This is a well-known and documented phenomenon. However, since shading and building materials that do not cause the heat island effect are readily available and economically viable, this kind of argument against density is considered unsubstantiated. Similarly, it is implicitly presumed that building best practices are employed for all new developments and retrofits to existing developments, to ensure that the collective lessons learned from the built environment are leveraged and that the same mistakes are not repeated.

3. Build In Passive Survivability (Immediate Action)

After the project has been sited, pursue opportunities that don't require active interventions to be successful.

- *Leverage open space* Where applicable, take advantage of the vast open spaces available in inland regions by designating natural stormwater retention ponds, green treatment, and, in severe cases, sacrificial land to protect settlements.
- *Promote crop diversity* In anticipation of increased average and extreme temperatures and storm events, as well as an increase or decrease in average rainfall, encourage and promote a greater spread in crop resiliency. Similarly, subsidize research and development for physical crop-protection systems.
- *Use water-efficient (xeric) plantings* This is not the "zero"-scape of rock lawns and bright concrete patios, but an attractive native and adapted landscape that is drought-tolerant and biologically active, and promotes local cooling through shade and evapotranspiration.
- *Build in passive cooling* Incorporate the multitude of measures that allow buildings to compensate for climatic extremes. These measures may include east–west building orientation, reflective roofing, thermally massive materials, reduced glazing areas (especially on west faces), exterior window shades, shaded arcades, and high-performance glazing.
- *Reduce heat islands* Through urban planning and building design, temper the formation of heat islands which artificially raise the urban temperature (see box "Density and Heat Islands"). Shade hardscapes through building form and selective use of xeric trees. Select materials with a heightened reflectivity (albedo) and emissivity (ability to reemit captured heat). Focus heat island reduction efforts at the urban scale using vulnerability mapping, a process that focuses support to vulnerable populations.

4. Design Active Resilient Systems (25+ years)

Once the passive opportunities have been exploited, design in climate-appropriate, low resource intensity systems that can withstand partial or complete disruptions.

- Building air-conditioning
  - *Consider compressor-based nonevaporative cooling* The energy efficiency advantage of evaporative systems is not as great as it once was. High efficiency compressor technology has advanced to close the gap.
  - *Select systems that have capacity to function in mixed modes (aka hybrid systems)* Such systems can more readily operate in a variety of modes ranging from "off," to "on with natural ventilation," to "on with air-conditioned supply." They do not require the relatively cold air (55°F [13°C]) and water (44°F [7°C]) typical of most air-conditioning systems, thus dampening the severity of system disruptions due to compromised power or water supply. Examples include radiant cooling, underfloor air, and displacement air systems.
  - *Leverage low nighttime temperatures* The 30°F+ (17°C+) day-to-night temperature variation is uniquely appropriate for thermal energy storage strategies (ice, chilled water, or condenser water).
  - *Wet/dry cooling towers* Where evaporative cooling towers are put to use, they should be hybrid towers capable of running the majority of the time in a dry "fluid cooler" mode and only occasionally in "wet" evaporative mode for extreme conditions. The water savings of these towers can exceed 95 percent of a conventional cooling tower.
  - *Adopt future design conditions* All systems should be specified in accordance with corrected climatic conditions appropriate to their system life (e.g., twenty-five years into the future for many heating, ventilation, and air-conditioning systems). The result may be slight increases to system size and lessened hours of natural ventilation modes during a typical year. Since many designers oversize systems already, a practice that can actually reduce performance, it is important that the system is right-sized for the future conditions, not simply further oversized. Systems that have a useful life of less than twenty-five years may not require this oversizing, but the space they are housed in or the paths they deliver energy in should be given that provision.
- Building plumbing
  - *Design water-efficient systems* Flow rates on fixtures should be no greater than 1.5 gallons per minute.[5] Flush rates on toilets and urinals should be no greater than 1.28 and 0.125 gallons per flush, respectively. Hot water latency should be less than three seconds (achievable through a variety of measures), and where possible foot-operated faucets should be installed to free up hands during washing.
  - *Reuse greywater* To use water only once and send it down the drain is a profligate waste of opportunity. So unless there is a regional water balance initiative (as there is in Las Vegas, Nevada), which requires return water flows to the local water treatment plant, passive greywater redirection to local landscape can be a valuable strategy.
- Build to lessen wildfire danger
  - *Gutters* Eliminate gutters or design to minimize risk.
  - *Vented roofs* Avoid vented roofs or protect vents from ember entry.
  - *Specify low-combustible building materials* Class A roofing, with noncombustible decking and siding.
  - *Manage planting* Through distancing, plant selection, and other strategies lessen the impacts of planting on fire exposure.
- Create resilient infrastructure systems (create a smart/resilient electrical and water networks that can suffer failures without ensuing cascading failures)
  - *Demand response* By shedding noncritical demands (electrical and water), the priority demands for life safety, economy, and environmental resilience can continue to be supplied in whole or in part.
  - *Diversify supply* By developing multiple supply sources (electrical and water), including the alternative supply "sources" of efficiency and closed-loop recovery, a failure in one supply can be compensated by another source.
  - *Informatics (making invisible information visible and visible information actionable)* By leveraging information technology and behavioral psychology, consumers and decision makers will have the opportunity to more effectively make decisions that enhance efficiency and cope with adversity.
- Plan resilient urban forms
  - *Corrected material selection* In addition to correcting the design day for thermal design, consider current and future species (such as termites) when selecting building materials.

– *Establish, expand, and optimize diverse modes of transportation in concert with transit-oriented development* Diverse transportation modes supported by appropriate urban density create improved access to critical community services in the event of failures. They also support vulnerable populations and encourage physical fitness. These benefits can help minimize heat illness risks due to climate change while simultaneously improving general health, culture, and economic strength in our communities.

5. Encourage Adaptable Buildings and Infrastructure (2+ to 200+ years)

Whatever decisions are made should be reasonably adaptable.

- *Design for how buildings learn[6]* By recognizing adaptability in layers, there is greater potential for building and infrastructure retrofits to be both lower cost and more effective. For example, plan renovations of older buildings to allow for easy retrofit with air-conditioning systems. And design lower-level wall systems for disassembly and ease of recladding if damaged.
- *Retrofit (i.e., teach) existing building stock* As our buildings age, capital reinvestment initiatives are an opportunity not only to beautify but to help the buildings better serve their occupants and owners. Examples include window retrofits, cool roofing, and ground-coupled air-conditioning.
- *Retrofit (i.e., teach) existing urban open space* To improve stormwater management and lessen urban heat islands, the street network should be gradually renewed with low-maintenance planting, best management practices, and partial inundation areas. Opportunities for stormwater capture and reuse should be explored.

6. Manage Settlement and Retreat (50+ years)

Although land management and water use practices continue to improve, desertification and water availability are threats to limitless growth. Adaptation plans in inland communities therefore need to make a reasonable effort to estimate their "right size" given available resources and scenario planning.

- *Managed settlement* As the carrying capacity of a region is approached, a variety of strategies may be used to manage growth while maintaining a strong economy. These include urban growth boundaries (à la Portland, Oregon), development impact fees (à la Albuquerque, New Mexico), and permit restrictions (à la Santa Fe, New Mexico's expired toilet credit program).
- *Passive retreat* As stresses increase, some retreat will occur through survival and/or market mechanisms. Commodity costs will increase. Water rights will be bought out by those willing to pay for them. This is likely to affect large water users (e.g., agriculture, golf courses, industrial users) most significantly.
- *Active retreat* See the box "Flooding in Pakistan." The authors believe that carefully planned active retreat will be needed in certain inland regions.

NOTES

1 United States Environmental Protection Agency, "A Student's Guide to Global Climate Change: Health," http://epa.gov/climatechange/kids/impacts/effects/health.html (accessed June 13, 2012).

2 Authors empirical analysis of World Bank Climate Change Portal data at http://sdwebx.worldbank.org/climateportal/index.cfm.

3 M. Carmen Moreno-Garcia, "Intensity and Form of the Urban Heat Island in Barcelona," *International Journal of Climatology* 14(6), 1993: 705–10.

4 "Urban Climate: Climate Study and UHI," United States Environmental Protection Agency, 2009.

5 Even at 1.5 gallons per minute, a typical shower will consume 15 gallons of water. In many modern global households (e.g., Delhi, India; Kyoto, Japan), middle-class citizens wash with less than three gallons of water by using hot water containers and ladles. It helps if the bathing room is warm.

6 Stewart Brand, *How Buildings Learn: What Happens After They're Built*, New York: Penguin Books, 1995.

Flooding in Pakistan

After the record rainfall and flooding in Pakistan in the monsoon season of 2010, there was mass movement of internally displaced persons (IDPs). A lot of the movement was toward temporary camps set up at the outskirts of large cities along the Indus River. IDPs included voluntary and early movers, but also those who were eventually actively resettled in temporary camps. Others who refused to abandon their homes despite increased flooding stress were injured and/or drowned.

Conversely, active resettlement (or active counter-retreat) was also witnessed. Though the emergency response and resettlement efforts were commendable overall, several nongovernmental organizations reported cases in which IDPs refused to return to their villages once water receded and conditions were determined to be safe—the life they had experienced in the temporary camps was of far better quality than they had ever had back home. Rumors of the government handing out free land and homes to IDPs were rampant and controlled just in time to avoid mass hysteria and revolt in camps outside Sukker, on the west bank of the Indus River. A large number of IDPs, mostly rural village farmers and fisherman, also moved to large cities such as Lahore, Quetta, and even as far south as Karachi, where they took up lives of homelessness, odd jobs, and street begging.

20.4 Internally displaced people in the aftermath of Pakistan's 2010 flooding
Image: Mahboob Khan, Arcop Associates

Glossary

In addition to supplemental definitions related to the planning, design, construction, and operation of the built environment, the authors have sourced definitions from leading scientific organizations. They include the Intergovernmental Panel on Climate Change (IPCC), the Center for Climate and Energy Solutions, the United States National Renewable Energy Laboratory, the United Kingdom Climate Impacts Program (UKCIP), and the World Bank. For the purposes of practice, many definitions have been adapted for application in the built environment and for ease of readability.

Accuracy. Degree of correctness, not repeatability. An analysis can be accurate but not precise, precise but not accurate, neither, or both.

Adaptation. Adjustment in natural or human systems to a new or changing environment. With regard to climate change, adaptation relates to adjustment in response to actual or anticipated climate effects. Adaptive measures may be anticipatory or reactive, initiated in the private or public sector, autonomous or planned.

Adaptation, integrated. Adaptation response that is holistic in approach and non-sector specific. A systemic response to adaptation that seeks development improvement, not just reduction in vulnerability.

Adaptive capacity. See *Coping ability*.

Additionality. The Kyoto Protocol requirement that greenhouse gas emissions after implementation of a project activity are lower than those that would have occurred in the most plausible alternative scenario to the project activity. If a project does not provide additional carbon emissions reductions compared to the plausible alternative scenario, then the emissions reductions cannot be traded.

Anthropogenic emissions. Emissions of greenhouse gases resulting from human activities.

Baselines. The baseline estimates of population, gross domestic product, energy use, and resultant greenhouse gas emissions without climate policies, used to determine the impacts of climate change policy.

Base year. Targets for reducing greenhouse gas emissions are often defined in relation to a base year. In the Kyoto Protocol, 1990 is the base year for most countries for the major greenhouse gases.

Basket of gases. The group of greenhouse gases regulated under the Kyoto Protocol. They are listed in Annex A of the Kyoto Protocol and include carbon dioxide (CO_2), methane (CH_4), nitrous oxide (N_2O), hydrofluorocarbons (HFCs), perfluorocarbons (PFCs), and sulfur hexafluoride (SF_6).

Carbon dioxide (CO_2). A colorless, odorless, nonpoisonous gas that is a normal part of air. Of the six greenhouse gases normally targeted, CO_2 contributes the most to human-induced global warming. CO_2 is the standard used to determine the "global warming potentials" of other gases. CO_2 has been assigned a one-hundred-year global warming potential of 1 (i.e., the warming effects over a one-hundred-year time frame relative to other gases).

Carbon dioxide equivalent (CO_2e). The emissions of a gas, by weight, multiplied by its global warming potential. CO_2e is used to compare the relative climate impact of different greenhouse gases. CO_2e allows reporting of greenhouse gas emissions using one standardized value and aids comparison of emissions generation or reduction.

Carbon emissions factor. The energy used at a building can be converted into carbon emissions using carbon emissions factors. These factors recognize that the amount of carbon emitted to produce 1 kilowatt hour of useful electricity is different from the amount of carbon emitted in the production of 1 kilowatt hour of useful heat from a gas boiler.

Carbon emissions intensity. A measure of the amount of CO_2e emissions per unit of gross domestic product. China and India have both proposed carbon emissions intensity reductions targets under the Copenhagen Accord.

Carbon neutral. A term used to describe a state in which net carbon emissions generated within a given emissions boundary equal zero. The measurement includes emissions related to energy used and energy exported, and any offsets that are purchased. Carbon neutrality is measured over a period of time, typically one year.

Carbon offset. A purchased or action credit that compensates for a carbon-equivalent emission.

Carbon sequestration. The removal of carbon from the atmosphere through processes such as storage of carbon in vegetation, or mechanical processes such as storage in geologic features.

Carbon sink. A process that removes carbon from the atmosphere.

Carbon tax. A surcharge on the carbon content of oil, coal, and gas.

Certified emissions reduction (CER). Reduction of greenhouse gas emissions achieved through a robust, traceable, additional process as required by the Kyoto Protocol. A CER can be sold or counted toward Annex I countries' emissions commitments. Reductions must be additional to any that would otherwise occur.

Chlorofluorocarbons (CFCs). Synthetic industrial gases composed of chlorine, fluorine, and carbon. They have been used as refrigerants, aerosol propellants, and cleaning solvents, and in the manufacture of plastic foam. There are no natural sources of CFCs.

Clean Development Mechanism (CDM). One of the three market mechanisms established by the Kyoto Protocol. The CDM is designed to promote sustainable development in developing countries and assist Annex I countries in meeting their greenhouse gas emissions reduction commitments. It enables industrialized countries to invest in emissions reduction projects in developing countries and to receive credits for reductions achieved.

Climate. The long-term average weather of a region, including typical weather patterns and the frequency and intensity of storms, cold spells, and heat waves. Climate is not the same as weather (see *Weather*).

Climate change. Changes in long-term trends in the average climate, such as changes in average temperatures. In IPCC usage, *climate change* refers to any change in climate over time, whether due to natural variability or as a result of human activity. In UNFCCC usage, climate change refers to a change in climate that is attributable directly or indirectly to human activity that alters atmospheric composition.

Climate-positive community. See *Net positive climate community*.

Climate sensitivity. The average global air surface temperature change resulting from a doubling of preindustrial atmospheric CO_2 concentrations. The IPCC estimates climate sensitivity at 1.5 to 4.5°C (2.7 to 8.1°F).

Climate variability. Changes in patterns, such as precipitation patterns, in the weather and climate.

Cogeneration. The use of a heat engine or power station to generate heat and electricity at the same time.

Commissioning (Cx). A systematic, documented quality control process that ensures that a building owner gets a building that operates as designed.

Concatenated risks. Risks that are connected or linked in a series, thereby resulting in risk accumulation.

Coping ability (adaptive capacity). Potential of a system or population to modify its features or behavior to cope better with existing and anticipated stresses. A sum of resistance and resilience: Coping ability = Resistance + Resilience.

Coping inability. The opposite of coping ability.

Cost and benefit analysis (CBA). A method for evaluating projects where the total cost of each option is compared to the total benefits.

Disaster. A situation or event that overwhelms local capacity.[1] A failure in governance and societal action to build adequate coping ability in advance of or in response to a significant event.

Discounting. The process that reduces future costs and benefits to reflect the time value of money and the common preference for consumption now rather than later.

Ecodistrict. A neighborhood or district with a broad commitment to accelerate neighborhood-scale sustainability. Systems synergy is the cornerstone of an ecodistrict, as one element in the district has a positive impact on another.

Ecosystem. A community of organisms and its physical environment.

Embodied carbon. The total equivalent carbon emissions associated with the processes of extraction, processing, delivery, and installation of a product.

Embodied emissions. The total emissions associated with the processes of extraction, processing, delivery, and installation of a product.

Embodied energy. The total energy associated with the processes of extraction, processing, delivery, and installation of a product.

Emissions. The release of substances (e.g., greenhouse gases) into the atmosphere.

Emissions boundary. A defined boundary typically used to calculate a carbon footprint or inventory. The boundary could be a building, organization, country, or person and a related activity, influence, or responsibility.

Emissions cap. A mandated restriction on the total amount of greenhouse gas emissions that an industry, country, or other entity can release into the atmosphere within a specific time span.

Emissions trading. A market mechanism that allows emitters (countries, companies, or facilities) to buy emissions from or sell emissions to other emitters. Emissions trading is intended to bring down the costs of meeting emission targets by allowing those who can achieve reductions less expensively to sell excess reductions (e.g., reductions in excess of those required under some regulation) to those for whom achieving reductions is more costly.

Energy conservation measures. Any project or process that will reduce the amount of energy used in a building.

Energy performance contracting. A turnkey service that provides customers with a comprehensive set of energy efficiency, renewable energy, and distributed generation measures. A typical energy performance contracting project is delivered by an energy service company and consists of the following elements: an initial audit through monitoring and verification, a comprehensive set of energy conservation measures, project financing and implementation, and a savings guarantee.

Evapotranspiration. The process by which water reenters the atmosphere through evaporation from the ground and transpiration by plants.

Externality. A positive or negative effect external to the option being evaluated and not accounted for in analysis.

General Circulation Model (GCM). A computer model of the basic dynamics and physics of the components of the global climate system (including the atmosphere and oceans) and their interactions, which can be used to simulate climate variability and change.

Global warming. The progressive gradual rise of the Earth's average surface temperature, thought to be caused in part by increased concentrations of greenhouse gases in the atmosphere.

Global warming potential (GWP). A system of multipliers devised to enable warming effects of different gases to be compared. The cumulative warming effect, over a specified time period, of an emission of a mass unit of CO_2 is assigned the value of 1.

Gray infrastructure. The network of sidewalks, roads, and piped systems used to manage the flow of stormwater and provide services to the built environment.

Greenhouse effect. The insulating effect of atmospheric greenhouse gases that keeps the Earth's temperature about 60°F (33°C) warmer than it would be otherwise.

Greenhouse gas (GHG). Any gas that contributes to the greenhouse effect.

Green infrastructure. The network of natural or constructed environmental features that manages the flow of stormwater, helps maintain healthy soils, and provides human benefits. Green infrastructure includes green spaces, street trees, green roofs, gardens, bioswales, and similar planted features.

Gross domestic product (GDP). A measure of the overall economic activity of a country.

Hazard. Any source of potential damage, harm, or adverse effects.

Hot air. A situation in which emissions (of a country, sector, company, or facility) are well below a target due to the target being above emissions that materialized in the normal course of events (i.e., without deliberate emission reduction efforts). Hot air can result from overoptimistic projections for growth. Emissions are often projected to grow roughly in proportion to GDP, and GDP is often projected to grow at historic rates. If a recession occurs and fuel use declines, emissions may be well below targets, since targets are generally set in relation to emission projections. If emission trading is allowed, an emitter could sell the difference between actual emissions and emission targets. Such emissions are considered hot air because they do not represent reductions on what would have occurred in the normal course of events.

Hydrofluorocarbons (HFCs). Synthetic industrial gases, primarily used in refrigeration and semiconductor manufacturing as commercial substitutes for chlorofluorocarbons (CFCs). There are no natural sources of HFCs.

Incentive-based regulation. A regulation that uses the economic behavior of firms and households to attain desired environmental goals. Incentive-based programs involve taxes on emissions or tradable emission permits. The primary strength of incentive-based regulation is the flexibility it provides the polluter to find the least costly way to reduce emissions.

Intangible. Something valuable that does not possess intrinsic productive value, e.g., trustworthiness.

Integrated Resource Management (IRM). A sustainability modeling and assessment tool created by Arup to support sustainable masterplanning and urban design activities. IRM provides the ability to test selected masterplan scenarios to derive an optimized solution through an iterative process of "assess-select-review-improve," and can be applied at different scales and levels of detail, ranging from individual developments to regions.

Intergenerational equity. The fairness of the distribution of the costs and benefits of a policy when costs and benefits are borne by different generations. In the case of a climate change policy, the impacts of inaction in the present will be felt in future generations.

Intergovernmental Panel on Climate Change (IPCC). The IPCC was established in 1988 by the World Meteorological Organization and the UN Environment Programme. The IPCC is responsible for providing the scientific and technical foundation for the United Nations Framework Convention on Climate Change (UNFCCC), primarily through the publication of periodic assessment reports.

Joint implementation. One of the three market mechanisms established by the Kyoto Protocol. Joint implementation is designed to allow developing countries to share greenhouse gas emissions reduction projects and sell the emissions reductions to developed nations.

Land use, land-use change, and forestry (LULUCF). One of the major categories in national greenhouse gas inventories prepared under the UNFCCC. It accounts for changes in the amount of biomass in a national system. An increase in biomass is considered to reduce greenhouse gas emissions, and a decrease in biomass is considered to increase greenhouse gas emissions. Deforestation is reported under LULUCF.

LEED. A green-building rating system developed by the U.S. Green Building Council.

Life-cycle cost analysis (LCCA). The analysis of economic costs over the period of an investment life. LCCA includes a variety of methods for determining if an alternative is economically justified, notably Life Cycle Costing (LCC), Net Savings (NS), Savings to Investment Ratio (SIR), Return on Investment (IRR or AIRR), Discounted Payback (DPB), and Simple Payback (SPB).

Maladaptation. Adaptation response that increases rather than reduces vulnerability (e.g., by transferring risks to other groups, future generations, or other ecosystems).

Methane (CH_4). CH_4 is among the six greenhouse gases to be curbed under the Kyoto Protocol. Atmospheric CH_4 is produced by natural processes, but there are also substantial emissions from human activities such as landfills, livestock and livestock wastes, natural gas and petroleum systems, coal mines, rice fields, and wastewater treatment.

Mitigation. Activities taken to reduce concentrations of atmospheric greenhouse gases and reduce the magnitude of future climate change.

Montreal Protocol. The Montreal Protocol on Substances that Deplete the Ozone Layer was designed to reduce the production and consumption of ozone-depleting substances in order to reduce their abundance in the atmosphere, and thereby protect the Earth's ozone layer. The original Montreal Protocol was agreed in 1987 and entered into force in 1989. It was the first UN treaty to achieve universal ratification, and it controls emissions of CFCs, HCFCs, and halons. These substances have global warming potential but are not covered by the Kyoto Protocol, as they were already covered by the Montreal Protocol.

Negative feedback. A process that results in a reduction in the response of a system to an external influence. For example, increased plant productivity in response to global warming would be a negative feedback on warming, because the additional growth would act as a carbon sink, reducing the atmospheric CO_2 concentration.

Net positive climate community (NPCC). A community that achieves net zero carbon community and goes further, to reverse historical carbon debt by sequestering carbon at an annually positive rate and/or measuring and reducing Scope 3 emissions using established accounting protocols.

Net zero carbon building (NZCB). A building that has greatly reduced *emissions* through emissions reductions, such that the balance of emissions for thermal and electrical energy within the building is met by on-site or off-site renewable energy or certified offsets. NZCBs are graded from A to D, based on their renewable energy supply and offset characteristics. They are equivalent to NZEBs accounted on an emissions basis.

Net zero carbon community (NZCC). A community that has greatly reduced *emissions* through emissions reductions, such that the balance of emissions for vehicles, thermal, and electrical energy within the community is met by on-site or off-site renewable energy or certified offsets. NZCCs are graded from A to D, based on their renewable energy supply and offset characteristics. They are equivalent to NZECs accounted on an emissions basis.

Net zero energy building (NZEB). A building that has greatly reduced *energy needs* through efficiency gains, such that the balance of energy for thermal and electrical energy within the building is met by on-site or off-site renewable energy. NZEBs are graded from A to D, based on their renewable energy supply characteristics. They account for energy on a site, source, cost, or emissions basis.

Net zero energy community (NZEC). A community that has greatly reduced *energy needs* through efficiency gains, such that the balance of energy for vehicles, thermal, and electrical energy within the community is met by on-site or off-site renewable energy. NZECs are graded from A to D, based on their renewable energy supply characteristics. They account for energy on a site, source, cost, or emissions basis.

Nitrous oxide (N_2O). N_2O is among the six greenhouse gases to be curbed under the Kyoto Protocol. N_2O is produced by natural processes, but there are also substantial emissions from human activities such as agriculture and fossil fuel combustion.

Nominal rate. An interest or escalation rate used in financial analysis that is inclusive of inflation.

Occurrence likelihood. The frequency with which or amount a system is exposed to a hazard. For the purposes of practice, the authors have used *occurrence likelihood* in lieu of *exposure* within the calculation of vulnerability.

Offset. A general term used to signify a purchase or other act that compensates for a use or emission. Offsets may be in the form of utility-sponsored green power purchasing, direct off-site renewable energy investments, or sponsored actions. The quality of offsets must be adequate to be considered acceptable for reporting and trading (e.g., certified carbon offsets, renewable energy certificates).

Operational energy. The amount of energy used by a development every year to run the development's systems. It includes electricity (from both grid and renewable sources) and all types of fuel used within the development.

Overshoot. The concept that the Earth's climate is "locked in" to a certain degree of change even if all greenhouse gas emissions stopped today. Overshoot describes the difference between the change that is observed and the change that will happen in the future as the climate system responds to the effect of increased greenhouse gases.

Parts per million/billion (ppm/ppb). The units in which concentrations of greenhouse gases are commonly presented. For example, since the preindustrial era, atmospheric concentrations of CO_2 have increased from 270 ppm to 370 ppm.

Perfluorocarbons (PFCs). PFCs are among the six types of greenhouse gases to be curbed under the Kyoto Protocol. PFCs are synthetic industrial gases generated as a by-product of aluminum smelting and uranium enrichment. They also are used as substitutes for CFCs in the manufacture of semiconductors.

Polluter Pays Principle (PPP). The principle that countries should in some way compensate others for the effects of pollution that they (or their citizens) generate or have generated.

Positive feedback. A process that results in an amplification of the response of a system to an external influence. For example, increased atmospheric water vapor in response to global warming would be a positive feedback on warming, because water vapor is a greenhouse gas.

Precision. Degree of repeatability, not correctness. An analysis can be accurate but not precise, precise but not accurate, neither, or both.

Radiative forcing. Net increases or decreases in the energy balance of the Earth–atmosphere system in response to greenhouse gases, land-use change, or solar radiation. Positive radiative forcings increase the temperature of the lower atmosphere, which in turn increases temperatures at the Earth's surface. Negative radiative forcings cool the lower atmosphere.

Real rate. An interest or escalation rate used in financial analysis that is exclusive of inflation.

Renewable Energy Certificates (RECs). Also known as Green Tags, Renewable Energy Credits, Renewable Electricity Certificates, or Tradable Renewable Certificates (TRCs), RECs are tradable, nontangible energy commodities that represent proof that electricity was generated from an eligible renewable energy resource.

Resilience. The ability to *recover* from the damage caused by a hazard. When paired with resistance, it is synonymous with coping ability (aka *adaptive capacity*). Coping ability = Resilience + Resistance. Note this is generally in keeping with IPCC TAR 2001 but differentiated from the UN/ISDR 2004

definition, which includes resistance. The authors separate out resistance for greater clarity in the practice of strategy development.

Resistance. The ability to *withstand* the damaging effect of a hazard. When paired with resilience, it is synonymous with coping ability (aka *adaptive capacity*). Coping ability = Resilience + Resistance.

Retrocommissioning (RCx). The application of the commissioning process to an existing building. It is usually used to realize energy savings or to correct building operations after several years of occupancy.

Risk. The product of the severity of a hazard and its occurrence likelihood. Risk does not include the effects of coping ability. Risk = Severity × Occurrence likelihood.

Sensitivity. The degree to which a system is affected, either adversely or beneficially, by climate variability or climate change. The effect may be direct (e.g., a change in crop yield in response to a change in the mean, range, or variability of temperature) or indirect (e.g., damages caused by an increase in the frequency of coastal flooding due to sea-level rise).

Sequestration. Opportunities to remove atmospheric CO_2 through either biological processes (e.g., plants and trees) or geological processes involving storage of CO_2 in underground reservoirs.

Severity. The degree to which a hazard is intense or impactful. For the purposes of practice, the authors have used *severity* in lieu of *sensitivity* within the calculation of vulnerability.

Site energy. The amount of energy that is used within the boundary of the development site. This includes metered electricity, all types of fuel, and energy from on-site renewables.

Smart grid. An electrical grid enabled with communications and measurement technology. This technology allows the grid operator and consumers to understand in more detail where power is being used, how savings can be made, and how loads are changing over time, which allows the grid operator to improve operations and consumers to improve use of energy.

Source energy. The primary energy used to extract, process, generate, and deliver energy to a site. To calculate a building's or community's total source energy, imported and exported energy is multiplied by the appropriate site-to-source conversion multipliers based on the utility's source energy type.

SPeAR. Sustainable Project Appraisal Routine. SPeAR is a holistic sustainability decision-making framework created by Arup to support project development and communicate outcomes. It is a software-based tool for use on a wide variety of projects and can be applied to design evaluation or project management. It encompasses an integrated quantitative and qualitative appraisal based on twenty-three core indicators.

Split incentive. Split incentives occur when the person paying the energy bills is not also making capital investment decisions. The most common form of split incentives is in leased buildings where tenants pay the energy bills, but owners pay for upgrades.

SRES scenarios. A suite of emissions scenarios developed by the IPCC in its Special Report on Emissions Scenarios (SRES). These scenarios were developed to explore a range of potential future greenhouse gas emissions pathways over the twenty-first century and their subsequent implications for global climate change.

Sulfate aerosols. Sulfur-based particles derived from emissions of sulfur dioxide (SO_2) from the burning of fossil fuels, particularly coal. Sulfate aerosols reflect incoming light from the Sun, shading and cooling the Earth's surface (see *Radiative forcing*), and thus offset some of the warming historically caused by greenhouse gases.

Sulfur hexafluoride (SF_6). SF_6 is among the six types of greenhouse gases to be curbed under the Kyoto Protocol. SF_6 is a synthetic industrial gas largely used in heavy industry to insulate high-voltage equipment and to assist in the manufacturing of cable-cooling systems.

Sustainability. To maintain, support, or endure. The most widely quoted definition of sustainability and sustainable development was given by the Brundtland Commission of the United Nations on March 20, 1987: "Sustainable development is development that meets the needs of the present without compromising the ability of future generations to meet their own needs."

Thermal expansion. Expansion of a substance as a result of the addition of heat. In the context of climate change, thermal expansion of the world's oceans in response to global warming is considered the predominant driver of current and future sea-level rise.

Thermohaline Circulation (THC). A three-dimensional pattern of ocean circulation driven by wind, heat, and salinity that is an important component of the ocean–atmosphere climate system. In the Atlantic, winds transport warm tropical surface water northward, where it cools, becomes more dense, and sinks into the deep ocean, at which point it reverses direction and migrates back to the tropics, where it eventually warms and returns to the surface.

Transit-oriented development (TOD). Development that is centered around public transit hubs and seeks to create walkable, compact communities. It is considered to be a key factor in reducing transportation carbon emissions.

Uncertainty. A prominent feature of the benefits and costs of climate change. Decision makers need to compare risk of premature or unnecessary actions with risk of failing to take actions that subsequently prove to be warranted. This is complicated by potential irreversibilities in climate impacts and long-term investments.

United Nations Framework Convention on Climate Change (UNFCCC). Treaty written in 1992 to cooperatively consider what countries could do to limit average global temperature increases and the resulting climate change, and to cope with whatever impacts were, by then, inevitable. The ultimate objective of the convention is to stabilize greenhouse gas concentrations "at a level that would prevent dangerous anthropogenic [human-induced] interference with the climate system."

Variable-air-volume (VAV) systems. Systems in which temperature control is provided by varying the amount of cool air provided to a room.

Vulnerability. The degree to which systems and populations are at risk and unable to cope with adverse impacts.[2] Vulnerability = Risk × Coping inability.

Water vapor (H_2O). The primary gas responsible for the greenhouse effect. It is believed that increases in temperature caused by anthropogenic emissions of greenhouse gases will increase the amount of water vapor in the atmosphere, resulting in additional warming (see *Positive feedback*).

Weather. The short-term (hourly and daily) state of the atmosphere. Weather is not the same as climate (see *Climate*).

NOTES

1   Definition adapted from that of the Centre for Research on the Epidemiology of Disasters.

2   Definition adapted from that of UKCIP.

Index

References in **bold** indicate tables, in *italics* indicate figures, and followed by a letter *n* indicate end-of-chapter notes.